Techniques for Building Timing-Predictable Embedded Systems

Nan Guan

Techniques for Building Timing-Predictable Embedded Systems

 Springer

Nan Guan
Assistant Professor
Department of Computing
Hong Kong Polytechnic University
Hung Hom, Kowloon, Hong Kong

ISBN 978-3-319-27196-5   ISBN 978-3-319-27198-9   (eBook)
DOI 10.1007/978-3-319-27198-9

Library of Congress Control Number: 2015959479

Springer Cham Heidelberg New York Dordrecht London

Printed on acid-free paper

Springer International Publishing AG Switzerland is part of Springer Science+Business Media (www.springer.com)

To Shiqing

Preface

Embedded systems are becoming more and more ubiquitous in our daily lives. Since the embedded systems interact with the physical environment, they often have to behave within specified time limits. We call such systems real-time systems. Violating the timing constraints may lead to disastrous consequences. Therefore, it must be guaranteed that the time constraints are satisfied in all situations. This is an extremely difficult task, as the number of possible system states is overwhelmingly large or even infinite.

In this book, we introduce some new methods to design and analyze real-time systems. Timing analysis of real-time systems is typically performed in a bottom-up manner, starting with the system's smallest components and then gradually up toward analyzing the system as a whole. The analysis on the program level aims to give each piece of program an upper limit of its execution time. On the component level (e.g., in a processor or a communication channel), many computation/communication tasks compete for the same platform, and the analysis should guarantee each of them obtains enough resource to process in time. On the system level, the analysis takes into account the interaction between the computation and communication activities that are distributed on various components. This book covers topics on each of these three levels.

On the program level (Part I) we study the worst-case execution time (WCET) analysis problem in the presence of caches with two commonly used replacement policies MRU and FIFO. Most of the research done with respect to the impact of caches to WCET of programs assume LRU caches. However, LRU is actually not commonly used in commercial processors because it requires a relatively complex hardware implementation. Hardware manufacturers tend to design the caches that are not using the LRU policy, particularly for embedded systems which are subject to strict restrictions in cost, power consumption, and heat. It has been found that existing methods of analysis, which is based on a qualitative classification of hits and misses in the cache, are not suitable to analyze MRU or FIFO cache. Our main contribution in this part is the development of a quantitative analysis methodology which is better suited to analyze the MRU and FIFO.

On the processor level (Part II) we address the challenges that arise due to the trend of multi-core processors. We first study some fundamental problems in multiprocessor real-time scheduling theory. Multiprocessor scheduling can be divided into two main paradigms, global and partitioned scheduling. In global scheduling, it is generally unknown what corresponds to the worst combination of contemporary software activations, which makes the analysis of such scheduling more difficult than its counterpart on single processors. Analysis of partitioned multiprocessor scheduling is simpler, but there is instead the challenge of how to allocate workload to individual processors for high resource utilization. We have studied the above theoretical problems and greatly advanced the state of the art. Moreover, we also consider practical aspects of real-time scheduling on multi-cores, to solve the interference arising between processor cores because of shared caches. We propose to use cache-aware scheduling where page-coloring are used to provide predictable cache performance of individual tasks.

On the system level, we introduce new techniques that solve the efficiency problem in the widely used real-time calculus (RTC) framework. Operations within the RTC generate curves with a frequency that is the product of the periodicity of the input curves. Therefore, RTC has an exponential complexity, and in practice the efficiency can be very low for complex systems. In this book, we present finitary real-time calculus to solve the above problems. The idea is to only maintain operations on the prefix of each trace which can affect the final result of the analysis. In that way we can show that analysis complexity is reduced to be pseudo-polynomial, and in practice the analysis efficiency is dramatically improved comparing with the ordinary RTC. An additional contribution to the RTC in this book is new analysis techniques of the earliest deadline first (EDF) scheduling algorithm in the RTC framework.

This book is the collection of some of the work I did during my Ph.D. study in Uppsala University, Sweden. I would like to thank my supervisor Wang Yi. I am truly grateful for his guidance, support, inspiration, patience, and optimism, especially when I was depressed by the negative aspects of my work as he consistently believes that there will be a way out and encourages me to carry on. I learned a lot more than computer science from him during the past 5 years. Without him, the work in this book would not have been possible to be finished (not possible to be started actually).

Kowloon, Hong Kong   Nan Guan
September 2015

Contents

Acronyms

AH	Always hit
AI	Abstract interpretation
AM	Always miss
BCET	Best-case execution time
CFG	Control flow graph
CHMC	Cache hit miss classification
D-PUB	Deflatable parametric utilization bound
DBF	Demand bound function
DM	Deadline monotonic
DRAM	Dynamic random-access memory
EDF	Earliest deadline first
FIFO	First-in-first-out
FM	First miss
FPS	Fixed-priority scheduling
FRTC	Finitary real-time calculus
GPC	Greedy processing component
HW	Hardware
IPET	Implicit path enumeration technique
ILP	Integer linear programming
LCM	Least common multiple
LHS	Left-hand side
LP	Linear programming
LRU	Least Recently Used
MBS	Maximal busy-period size
MILP	Mixed integer linear programming
MPSoC	Multiprocessor system on chip
MRU	Most Recently Used bit
NC	Nonclassified
NP-EDF	Non-preemptive earliest deadline first
NP-FP	Non-preemptive fixed-priority
P-FP	Preemptive fixed-priority

PLRU	Pseudo-Least Recently Used
PUB	Parametric utilization bound
RHS	Right-hand side
RM	Rate monotonic
RMS	Rate monotonic scheduling
RTA	Response time analysis
RTC	Real-time calculus
SBF	Service-bound function
SCC	Strongly connected component
SW	Software
WCET	Worst-case execution time
WCRT	Worst-case response time

Chapter 1
Introduction

Embedded systems are playing a more and more important role in our daily life; they exist everywhere in our society from consumer electronics such as multimedia systems, mobile phones, microwave ovens, refrigerators to industrial fields such as robotics, telecommunications, environment monitoring, and nuclear power plants. According to statistics reported by ARTEMIS [1], 98 % computing devices in the world are embedded systems.

Due to close interaction with physical world, embedded systems are typically subject to timing constraints. They are often referred as to *real-time systems*. Violating timing constraints is fatal to such a system, which may lead to catastrophic consequences such as loss of human life.

Therefore, at design time, it must be guaranteed that the system satisfies the pre-specified timing constraints at run-time under any circumstance. For all but very simple systems, the number of possible execution scenarios is either infinite or excessively large, where exhaustive testing cannot be used to verify the timing correctness. Instead, formal analysis techniques are necessary to ensure the timing predictability.

To meet the increasing requirements on functionality and quality of service, embedded systems are becoming more powerful and more cooperative, moving from dedicated single-functionality control logics to complex systems hosting multiple tasks executing concurrently and distributively over a networked platform. Timing analysis of such systems is typically performed in a bottom-up manner, consisting of:

- *Program-level analysis*. Timing analysis is first performed on the program level, to bound the worst-case execution time (WCET) of each individual task assuming fully dedicated hardware.
- *Component-level analysis*. Then the WCET information of individual tasks on a component (a processing unit or a communication interconnection) is gathered, to investigate the contention of processing capacity among different tasks in a multitasking environment.

© Springer International Publishing Switzerland 2016
N. Guan, *Techniques for Building Timing-Predictable Embedded Systems*,
DOI 10.1007/978-3-319-27198-9_1

- *System-level analysis.* Finally, the analysis is conducted in the scope of the whole network, considering the interaction among computation and communication activities distributed over different processing units and interconnections.

Timing analysis of a computer system has to take into account the underlying hardware platform. Modern processors rely on fast caches, deep pipelines, and many speculative mechanisms such as prefetching to achieve high computation power. The product of these complex hardware features and the inherent complexity of software programs make it a very difficult problem to efficiently and precisely predict the WCET of even a small piece of code. For example, minor mis-predictions of the cache access behavior may cause great deviation in the analysis of a program's WCET, since the execution time of individual instructions may vary by several orders of magnitude depending on whether the memory access is a hit or miss on the cache.

A more significant trend in computer hardware is the paradigm shift to multi-core processors. The processor manufacturers have run out of room to boost processor performance with their traditional approaches of driving clock frequency further and further. Instead, processors are evolving in the direction of integrating more and more processing cores on a single chip, which provides higher computation power at lower cost of energy consumption. This brings great opportunities to embedded systems area, and at the same time also new challenges. While the scheduling of tasks on traditional uniprocessor platforms only involves a single dimension in time, i.e., to decide *when* to execute a certain task, the problem becomes two dimensional on multi-core processors as it also needs to decide *where* (i.e., on which core) to execute the task. Apart from the processing cores, different tasks also contend on many other resources in multi-core processors, such as shared caches, shared buses, and shared memory interfaces. Interleaving of concurrent accesses to these shared resources creates a tremendous state space of the system behavior, making its timing analysis extremely difficult if not impossible at all [2].

The number of cores integrated in a multi-core chip is increasing rapidly. Chips with hundreds of cores are not rare in the market today, and the industry has recently coined the "New Moore's law," predicting that the number of cores per chip will double every 2 years over the next decade [3]. On the other hand, heterogenous multi-core architectures using dedicated cores for specific functionalities yield simpler structures, less power consumption, and higher performance. Therefore, the design and analysis of real-time systems on future multi-core processors will be founded on a large-scale heterogenous networking base, facing more challenging scalability and diversity problems than traditional macro-world distributed real-time systems.

The aim of this thesis is to develop new techniques to address several important problems raised by above challenges in the design and analysis of real-time embedded systems, in particular covering the following aspects:

- On the program level, we study the problem of predicting cache access behaviors to precisely estimate the WCET of a program.

- On the component level, we investigate design and analysis techniques for scheduling real-time applications on multi-core processors.
- On the system level, we improve both the scalability and precision of Real-Time Calculus (RTC) for modular performance and timing analysis of distributed real-time systems.

1.1 Technical Background

This chapter presents a brief introduction to the technical areas addressed in this thesis.

1.1.1 WCET Estimation and Cache Analysis

The execution time of a program depends on the inputs and hardware states. It is usually infeasible to determine the exact WCET of a program due to the huge number of inputs and initial hardware states. The goal of WCET estimation [4] is to compute safe upper bounds of, and preferably close to, the real WCET value of the program. Figure 1.1 shows the main building blocks of WCET estimation:

- *Control-flow Reconstruction* analyzes and disassembles the binary executable, and transforms it into a control-flow graph (CFG) [5].
- *Program-behavior Analysis* annotates the CFG with extra information to further regulate its possible behaviors, e.g., bounding the maximal iterations of loops (loop analysis) [6–9] and excluding infeasible paths (control-flow analysis) [7, 10–12].
- *Micro-architectural analysis* determines the execution delay of each program fragment, e.g. each instruction or each basic block. This needs to consider the effect of hardware components, e.g., pipelines [13–16], caches, branch prediction, memory controllers, etc.
- *Path Analysis* determines the execution time bounds of the overall program, using the output of micro-architectural analysis and the annotated CFG [17–20].

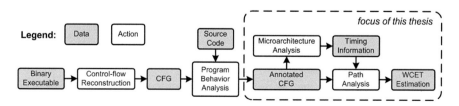

Fig. 1.1 Main building blocks of WCET estimation

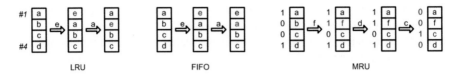

Fig. 1.2 Illustration of the LUR, FIFO and MRU cache replacement policies

The focus of this thesis is on the scope of micro-architectural analysis and path analysis. In particular, we consider the *cache* issue in micro-architectural and path analysis.

A *cache* is a small, fast memory which stores copies of frequently used data in the main memory to reduce memory access latencies. At run-time, when the program accesses the memory, the processor first checks whether the content is in the cache. If yes, it is a *hit*, and the content is directly accessed from the cache. Otherwise, it is a *miss*, and the content is installed in the cache while the program accesses it from the memory. The cache is usually too small to accommodate all the memory contents needed by a program, so a *replacement policy* must decide which memory content to be evicted from the cache upon a miss.

Commonly used cache replacement policies include LRU (least-recently-used) [19], FIFO (first-in-first-out) [21], MRU (most-recently-used-bit) [22], and PLRU (pseudo-least-recently-used) [23]. Figure 1.2 depicts the replacement rules of LRU, FIFO and MRU. LRU always stores the most recently accessed memory block in the first cache line. FIFO behaves in the same way as LRU upon a miss, but does not change the cache state upon a hit. MRU stores an extra MRU-bit, being 1 indicates that this line was recently visited. When the program accesses a memory content s, MRU first checks whether s is already in the cache. If yes, then s is still stored in the same cache line and its MRU-bit is set to 1 regardless of its original state. If s is not in the cache, MRU finds the first cache line whose MRU-bit is 0, then replace the originally stored memory block in it by s and set its MRU-bit to 1. After the above operations, if there still exists some MRU-bit at 0, the remaining cache lines' states are kept unchanged. Otherwise, all the other cache lines' MRU-bits are changed from 1 to 0.

Exact characterization of cache access behaviors in WCET estimation suffers from serious state-space explosion. The state-of-the-art techniques are based on an over-approximate analysis framework consisting of cache analysis by abstract interpretation (AI) [24] and path analysis by implicit path enumeration [18]. The idea is to derive an upper bound for the timing delay of each program segment (whenever it is executed) such that the WCET bound is obtained by finding the weighted longest path in the CFG, which can be efficiently solved as an integer linear programming problem.

The main target in AI-based cache analysis is to decide the cache hit/miss classification for each program point [24, 25]:

AH (always hit): The memory accesses are always hits whenever the program point is visited.

FM (first miss): The memory accesses are a miss for the first execution, but always hits afterwards. This classification is useful to handle "cold miss" in loops.

AM (always miss): The memory accesses are always misses whenever the program point is visited.

NC (non-classified): The memory accesses cannot be classified into any of the above categories. This category has to be treated as AM in the path analysis.

These four classifications can be grouped into two categories: AH and FM are *positive* since they ensure that (the major portion of) the memory accesses of a node to be hits, while AM and NC are *negative* since any program point belong to these classifications are treated as always miss in the path analysis.

The analysis maintains a set of possible cache states at each program point, and iteratively updates the cache states according to the control flow and memory access at each point. This procedure terminates as soon as the cache states at all program points reach fixed-points, i.e., no new possible cache states are generated if the update procedure continues. Now the fixed-point includes all the possible cache states at each program point, by which the cache hit/miss classification is determined. For example, the memory content at a program point is AH if it appears in all possible cache states when the fixed-point is reached.

The number of possible cache states at each program point may explode exponentially during the procedure described above. For scalability, different possible cache states can be merged into certain form of abstractions, with the update operations performed in the abstract domain [24]. Constructing the abstraction domain is specific to a certain replacement policy, and it preferably maintains as much as possible information by a limited amount of data, to get both good precision and scalability.

In general, LRU is considered to be much more predictable than other commonly used replacement policies [26]. LRU is recommended as the only option for the cache replacement policy when the timing predictability is a major concern in system design [27], and most research work in real-time systems involving cache issues assumes LRU caches by default. There have been few works on the analysis of non-LRU replacement policies, which end up with either considerably lower precision due to the difficulty of constructing precise abstract domains [21, 23, 28] or poor scalability due to state-space explosion in explicit cache state enumeration [29].

On the other hand, many non-LRU replacement policies enjoy the benefit of simpler hardware implementations and are preferred than LRU in practical processor design. This raises an important yet challenging problem of how to precisely analyze these non-LRU cache replacement policies for real-time systems. As the existing analysis framework has shown inadaptability to non-LRU caches, new insights from a different perspective would be necessary to address this problem. In the next chapter, we will present a novel *quantitative* cache analysis method and use it to precisely predict the cache/hiss behavior of MRU and FIFO caches for tighter WCET estimation.

1.1.2   Multiprocessor Real-Time Scheduling

The aim of *real-time scheduling* is to execute multiple applications on a shared hardware platform such that their timing constraints are always met. The software system is modeled by a *sporadic* task set τ [30], which consists of N independent tasks. Each sporadic task τ_i is characterized by a tuple (C_i, D_i, T_i), where C_i is the WCET, D_i is the relative deadline, and T_i is the minimum inter-arrival separation time (also referred to as the period) of the task. We further define the utilization of a task as $U_i = \frac{C_i}{T_i}$, which represents the portion of the processing capacity requested by this task in the long term. In general, the relative deadline D_i of a task may be different from its period T_i. An *implicit-deadline* task τ_i satisfies the restriction $D_i = T_i$, a *constrained-deadline* task τ_i satisfies $D_i \leq T_i$, whereas an *arbitrary-deadline* task τ_i does not limit the relation between D_i and T_i, and in particular allows $D_i > T_i$.

A sporadic task generates a potentially infinite sequence of jobs with successive job-arrivals separated by at least T_i time units. A job released by task τ_i at time instant r needs to finish its execution (for up to C_i time units) no later than its absolute deadline $d = r + D_i$. The *response time* of a job is the timing delay between its release and time point when it is finished. The *worst-case response time* (WCRT) [31] R_i of task τ_i is the maximal response time value among all jobs of τ_i in all job sequences possible in the system. $S_i = D_i - R_i$ is the worst-case slack time of task τ_i.

Fixed-priority scheduling (FPS) [32] is one of the most commonly used approaches to schedule real-time tasks. In FPS, each task is assigned a priority and all jobs released by the task inherit this priority. The active jobs in the ready queue are ordered by their priorities, and the scheduler always picks the job at the head of the ready queue (i.e., with the highest priority) for execution. We follow the convention in most real-time scheduling literatures that tasks are ordered by their priorities and a smaller index implies a higher priority.

The scheduler can be either preemptive or non-preemptive. A preemptive scheduler may re-order the ready queue as soon as some job is released, such that it may suspend the currently executing job and executes the newly release job with a higher priority. In a non-preemptive scheduler, the ready queue is re-ordered only at job completion, and thus each job occupies the processor until completion without interference once it starts execution.

When $D_i > T_i$, it is possible that several jobs of a task are active simultaneously. We restrict that a job can execute only if its precedent job has been already finished, which avoids unnecessary working space conflict and automatically resolves the data dependencies among jobs of the same task. This restriction is commonly adopted in the implementation of real-time operating systems, e.g., RTEMS [33] and LITMUS^{RT} [34].

An important problem in embedded systems design is to analyze the *schedulability* of the task set. A task is schedulable iff all of its released jobs can finish before their absolute deadlines, and a task set τ is schedulable iff all tasks in τ are schedulable. The analysis of sporadic task sets preemptively scheduled on *uniprocessor* platforms has been well studied since the seminal work by Liu and Layland [32]. An essential concept of its analysis is the *critical instant*, which describes the release time of a task for which its response time is maximized. Liu and Layland proved that [32]:

> A critical instant for any task occurs whenever the task is requested simultaneously with requests for all higher priority tasks.

Therefore, the schedulability analysis can focus on one concrete release pattern, despite the infinitely many release scenarios that are possible at run-time. Based on this worst-case release pattern, one can calculate the worst-case response time by (symbolically) simulating the execution sequence (response time analysis [31, 35]), and compare it with the relative deadline to verify a task's schedulability. The knowledge of critical instant forms the foundation of uniprocessor fixed-priority schedulability analysis for sporadic task systems and many of its variants, with non-preemptive blocking [36, 37], resource sharing [38], release jitters and burst behaviors [39, 40], offsets between tasks [41, 42], task dependencies [43, 44], etc.

Based on the critical instant, Liu and Layland developed another classical result in uniprocessor FPS: the *rate-monotonic* (RM) priority assignment policy (tasks with shorter periods have higher priorities) is optimal for implicit-deadline task systems, with which the schedulability is guaranteed by the following condition:

$$\sum_{\forall \tau_i \in \tau} U_i \leq N \times (2^{\frac{1}{N}} - 1)$$

The expression $N \times (2^{\frac{1}{N}} - 1)$ is known as *Liu and Layland's utilization bound*, which is a monotonically decreasing function with respect to N, the number of tasks, and reaches its minimum $\ln(2) \approx 0.693$ as N approaches infinity. Liu and Layland's utilization bound gives a very simple *sufficient* schedulability test, which can be used for, e.g., admission control and workload adjustment at run-time [45, 46].

The above results are all for the case that tasks are scheduled on a uniprocessor. However, the scheduling problem becomes much more difficult if the execution platform consists of multiple processors, as noted by Liu [47]:

> Few of the results obtained for a single processor generalize directly to the multiple processor case ... The simple fact that a task can use only one processor even when several processors are free at the same time adds a surprising amount of difficulty to the scheduling of multiple processors.

Multiprocessor scheduling can be categorized into two major paradigms [48, 49]: *global scheduling*, in which each task can execute on any available processor at run-time, and *partitioned scheduling* in which each task is assigned to a processor beforehand, and at run-time each task can only execute on this particular processor.

Global scheduling [50–55] on average utilizes computing resource better, and is more robust in the presence of timing errors. A major obstacle in analyzing global scheduling is the unknown critical instant. The critical instant of fixed-priority uniprocessor scheduling (simultaneous releases of higher priority tasks) does not necessarily lead to the worst-case response time in global scheduling [56, 57]. Moreover, global scheduling algorithms based on widely optimal uniprocessor scheduling algorithms like RM and EDF (earliest-deadline-first) [32] suffer from the so-called *Dhall effect* [58], namely some system with total utilization arbitrarily close to 1 may be infeasible by global RM/EDF scheduling no matter how many processors are added to the system.

Partitioned scheduling [58–64] does not allow dynamic load balancing, but enjoys the benefit of relatively easier analysis. As soon as the system has been partitioned into subsystems that will be executed on individual processors, the uniprocessor real-time scheduling and analysis techniques can be applied to each subsystem. Workload partitioning to individual processors is similar to the well-known intractable bin-packing problem [65]. Theoretically, the worst-case utilization bound of partitioned scheduling cannot exceed $50\% \times M$ (M is the number of processors) regardless of the local scheduling algorithm on each processor [48]. In order to break through this hard limitation and get higher utilization bounds, task splitting mechanisms are added to standard partitioned scheduling, such that a small number of tasks are allowed to migrate among cores under strict control [66–69].

The multiprocessor real-time scheduling problem has drawn a rapidly increasing interest during the last decade, motivated by the significant trend of *multi-core* processors. Typical multi-core architectures integrate several cores on a single processor chip. The cores usually share a memory hierarchy including L2/L3 caches and DRAM, and an interconnection infrastructure offering communication mechanism among them. Thanks to these on-chip shared resources, the cost of task migration is greatly reduced and thus scheduling algorithms allowing task migration (e.g., global scheduling) become practical options. However, on-chip shared resources also cause serious negative effects to timing predictability of the system: the timing behavior of a task depends on co-running tasks due to non-deterministic contentions on shared resources, which invalidates the traditional analysis framework with independent program-level and component/system-level timing analysis. This problem is a major obstacle to use multi-core processors in real-time embedded applications [70, 71].

Today, 40 years after the publication of Liu and Layland's seminal paper [32], people have gained good understanding of how to design and analyze real-time systems on uniprocessor platforms. However, the field is far from mature for multi-core systems. From the fundamental multiprocessor scheduling theory to practical solutions to build efficient and predictable real-time embedded systems are still open. In the next chapter, we will address some of these important problems, extending Liu and Layland's classical results in uniprocessor scheduling theory to multiprocessor scheduling and solving the shared cache contention problem for real-time scheduling on multi-core processors.

1.2   Real-Time Calculus

Many embedded systems have functionalities distributed over multiple processing components and communicated by different interconnections. One method to analyze the timing properties of distributed systems is the *holistic* approach [72, 73], which extends the classical analysis techniques in uniprocessor scheduling to certain workload models and resource arbitration policies in a distributed computing environment. The holistic approach allows to take global dependencies into account, which benefits in analysis precision. However, it suffers relatively higher analysis complexity, and needs to be adopted to specific workload and resource arbitration models.

A different way for timing analysis of distributed systems is the *modular* approach [74, 75], which divides the analysis of the whole systems into separate analysis on local components and integrates individual local analysis results to obtain performance characterizations of the complete system. Comparing with the holistic approach, the modular approach exhibits significantly better flexibility and scalability, and thus is prevailing in the analysis of complex distributed embedded real-time systems.

This thesis considers a specific modular performance analysis approach called RTC [76], which originates from the Network Calculus [77] theory for deterministic traffic analysis of network systems. In RTC, workload and resource availability is modeled using variability characterization curves [78], which generalize many existing task and resource models in real-time scheduling theory. RTC has proved to be one of the most powerful methods for real-time embedded system performance analysis, and has drawn a lot of attention in recent years.

In RTC, the workload is modeled by *arrival curves*. Let $R[s, t)$ denote the total amount of requested capacity to process in time interval $[s, t)$. Then, the corresponding upper and lower arrival curves are denoted as α^u and α^l, respectively, and satisfy:

$$\forall s < t, \ \alpha^l(t - s) \leq R[s, t) \leq \alpha^u(t - s) \tag{1.1}$$

where α^u and α^l are functions with respect to time interval sizes and $\alpha^u(0) = \alpha^l(0) = 0$.

The resource availability is modeled by *service curves* in RTC. Let $C[s, t)$ denote the number of events that a resource can process in time interval $[s, t)$. Then, the corresponding upper and lower service curves are denoted as β^u and β^l, respectively, and satisfy:

$$\forall s < t, \quad \beta^l(t - s) \leq C[s, t) \leq \beta^u(t - s) \tag{1.2}$$

where β^u and β^l are functions with respect to time interval sizes and $\beta^u(0) = \beta^l(0) = 0$.

A software processing system or a dedicated HW unit is modeled as a *component*. A workload stream represented by arrival curves α^u and α^l enters the component and is processed with resource represented by service curves β^u and β^l. The component generates output workload stream represented by arrival curves $\alpha^{u'}$ and $\alpha^{l'}$, and the remaining resources is represented by service curves $\beta^{u'}$ and $\beta^{l'}$.

In RTC we can model HW/SW processing units with different resource arbitration policies. An example is the *Greedy Processing Component* (GPC), which processes computation requests from the input workload stream in a greedy fashion. The arrival and service curves that characterize the output of a GPC is computed by [74]:

$$\alpha^{u'} \triangleq \min((\alpha^u \otimes \beta^u) \oslash \beta^l, \beta^u) \tag{1.3}$$

$$\alpha^{l'} \triangleq \min((\alpha^l \oslash \beta^u) \otimes \beta^l, \beta^l) \tag{1.4}$$

$$\beta^{u'} \triangleq (\beta^u - \alpha^l) \overline{\oslash} 0 \tag{1.5}$$

$$\beta^{l'} \triangleq (\beta^l - \alpha^u) \overline{\otimes} 0 \tag{1.6}$$

where the min-plus convolution $\otimes$, max-plus convolution $\overline{\otimes}$, min-plus deconvolution $\oslash$, and max-plus deconvolution $\overline{\oslash}$ are defined as:

$$(f \otimes g)(\Delta) \triangleq \inf_{0 \leq \lambda \leq \Delta} \{f(\Delta - \lambda) + g(\lambda)\} \tag{1.7}$$

$$(f \overline{\otimes} g)(\Delta) \triangleq \sup_{0 \leq \lambda \leq \Delta} \{f(\Delta - \lambda) + g(\lambda)\} \tag{1.8}$$

$$(f \oslash g)(\Delta) \triangleq \sup_{\lambda \geq 0} \{f(\Delta + \lambda) - g(\lambda)\} \tag{1.9}$$

$$(f \overline{\oslash} g)(\Delta) \triangleq \inf_{\lambda \geq 0} \{f(\Delta + \lambda) - g(\lambda)\} \tag{1.10}$$

We can model the FPS by cascading several GPC with their resource inputs and outputs connected. For different resource arbitration policies, tailored relations between the input and output curves need to be established. Under some policies such as FIFO and EDF, the calculation of an output workload stream may need information of all other streams involved in the resource arbitration, which can be modeled by components with multiple pairs of input and output workload streams.

When a workload stream with upper arrival curve α^u is processed by a GPC on a resource with lower service curve β^l, the maximal *delay* to complete a computational request issued in the workload stream is bounded by [74]:

$$D(\alpha^u, \beta^l) \triangleq \sup_{\lambda \geq 0} \left\{ \inf\{\tau \in [0, \lambda] : \alpha^u(\lambda - \tau) \leq \beta^l(\lambda)\} \right\}$$

On the other hand, we can bound the maximal *backlog*, i.e., the total amount of unfinished workload that has been requested at any time instant, which is useful, e.g., to determine the buffer size to store unprocessed input data [74]:

$$B(\alpha^u, \beta^l) \triangleq \sup_{\lambda \geq 0} \left\{ \alpha^u(\lambda) - \beta^l(\lambda) \right\}$$

Intuitively, $B(\alpha^u, \beta^l)$ and $D(\alpha^u, \beta^l)$ are the maximal vertical and horizontal distance from α^u to β^l.

The RTC framework connects multiple components into a network to model systems with networked structures. The analysis of the whole network follows the workload and resource flows, starting with a number of initial input curves and generating output curves step by step to traverse all the components in the network. In case the network contains cyclic flows, the analysis starts with under-approximation of the initial inputs and iterates until a fixed point is reached [79]. RTC Toolbox [80] is an open source Matlab library that implements the state-of-the-art modeling capacity and analysis techniques of the RTC framework.

Although RTC has proved itself a successful framework for the analysis of many distributed real-time systems, some major limitations need to be addressed to extend its competitiveness to a wider scope. On one hand, RTC may run into serious scalability problems for large-scale systems with complex timing characterizations. On the other hand, the analysis precision is still unsatisfactory for components with complex arbitration policies such as EDF. Both problems will be addressed in the next chapter, with new techniques exploring implicit useful information that has not been investigated in the original RTC framework.

1.3 What's New in This Book

In this chapter we briefly describe the main contributions of this thesis in each of the areas presented above.

1.3.1 Quantitative Cache Analysis for WCET Estimation

Although LRU holds a dominating position in research of real-time systems, it is actually not so common to see processors equipped with LRU caches on the market. The reason is that the hardware implementation of LRU is relatively expensive [81].

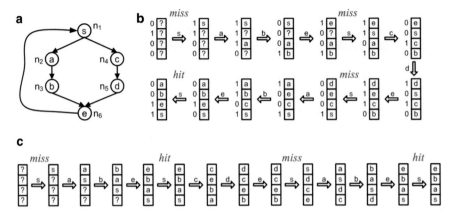

Fig. 1.3 An example to illustrate the cache access behavior under MRU and FIFO

On the other hand, many non-LRU replacement policies enjoy simpler implementation logic, but still have almost as good average-case performance as LRU [82].
Therefore, hardware manufacturers tend to choose these non-LRU replacement
policies, especially for embedded systems that are subject to strict cost, power, and
thermal constraints.

In this thesis, we develop a quantitative cache analysis approach and apply it to
two common non-LRU replacement policies, MRU (Chap. 2) and FIFO (Chap. 3).
MRU has been used in mainstream processor architectures like Intel Nehalem [83]
and UltraSPARC T2 [84], and FIFO is adopted in many processor series of Intel
XScale, ARM9, and ARM11 [26]. The existing cache analysis techniques based
on *qualitative* cache hit/miss classifications (either positive or negative) are not
adequate to precisely capture the cache miss/hit behavior under MRU and FIFO
replacement, as shown in the following example.

Figure 1.3a shows the CFG of a loop structure, and we consider a particular
execution sequence in which the two branches are taken alternatively. Figure 1.3b, c
depict the cache state update under MRU and FIFO replacement, respectively. Under
MRU, the first three accesses to s are all misses, but the fourth access is a hit. If the
sequence continues, the accesses to s will always be hits afterwards. Under FIFO,
hits and misses occur alternatively. In both cases, the first miss is a cold miss which
is unavoidable under our initial cache state assumption, but the other misses are all
because s is evicted by other memory blocks. Indeed, node n_1 cannot be determined
as AH or FM, and one has to put it into the NC classification and treat it as being
always a miss whenever it is executed. This shows the inherent limitation of the
existing cache analysis techniques based on *qualitative* cache access classifications,
which treat each node as either "black" or "white." But in fact many nodes of the
program has a "grey" behavior. For a safe approximation they have to be treated
as "black," which is inherently very pessimistic in predicting the overall execution
time of the program.

In order to precisely characterize the "grey" behaviors, we introduce *quantitative*
cache access classifications. In contrast to the qualitative classifications, quantitative

classifications bound the number of misses incurred at individual or a set of program points, hence can more precisely capture the nuanced behavior of MRU and FIFO caches. In particular, we introduce the k-miss classification for MRU which states that at most k accesses are misses while all the others are hits, and introduce the γ-set classification for FIFO which bounds the ratio of misses out of all memory accesses of certain program points. In the above example, the cache access behavior of node s can be captured by the 3-miss classification and $\frac{1}{2}$-set classification under MRU and FIFO, respectively.

It turns out that it is a difficult problem to directly decide whether a program point can be put into these quantitative classifications in the analysis framework of abstract interpretation. Our solution is to reduce the verification of quantitative properties to qualitative problems. We use the cache analysis results of the same program under LRU to derive k-Miss classification under MRU, which inherits the advantages in both efficiency and precision from the state-of-the-art LRU analysis based on abstract interpretation [19].

Experiments with benchmark programs show that the proposed quantitative cache analysis methods have both good precision and efficiency. For both MRU and FIFO caches under typical settings, the over-estimation ratio of the WCET obtained by our quantitative analysis method is around 10 %, and the analysis procedure for each benchmark program terminates in seconds. These results also suggest that the predictability of MRU and FIFO has been under-estimated before. They may still be reasonable candidates of cache replacement policies for real-time systems, considering their other advantages in hardware cost, power consumption, and thermal output.

1.3.2 Real-Time Scheduling for Multi-Cores

The aim of this part is to gain better understanding of the design and analysis principle of real-time systems on multi-core platforms. Although the uniprocessor real-time scheduling theory has been well established, many important problems are still open in the multi-core setting. In this thesis we look into several fundamental problems in multiprocessors scheduling theory, with both the global (Chaps. 4 and 5) and partition-based (Chaps. 6 and 7) approaches. Moreover, in Chap. 8 we consider the unpredictable inter-core interferences caused by contentions on shared caches.

1.3.2.1 Global Scheduling

As mentioned in Sect. 1.1.2, the critical instant in global FPS is in general unknown. The simultaneous release of higher-priority tasks does not necessarily lead to the worst-case response time in global scheduling. Existing analysis techniques either (explicitly or symbolically) enumerate all the possible system behaviors, which is

extremely inefficient due to state-space explosion [85–87], or safely cover all the
possible behaviors by approximating the system workload, which results in over-
pessimistic analysis results.

The first contribution of this part is to deal with the problem of unknown
critical instant in global scheduling. Chapter 4 develops the concept of *abstract
critical instant* for the analysis of fixed-priority global scheduling, where all higher-
priority tasks, except $M - 1$ of them (M is the number of processors), are requested
simultaneously with the task in question. Even though the abstract critical instant
still does not provide precise information about the worst-case release times of the
higher priority tasks, we are left with a set among which the real critical instant
can be found. This set is significantly smaller than the whole space of all possible
job sequences, which enables very precise analysis by excluding impossible system
behaviors from the calculation of interference. In Chap. 5 we extend the above
insights to *non-preemptive* global FPS, by counting the blocking of low-priority
tasks, and propose several schedulability test conditions to compromise the analysis
efficiency and precision.

Another important difference between fixed-priority global scheduling and its
uniprocessor counterpart is the condition for bounded responsiveness. In fixed-
priority uniprocessor scheduling, each task of a task set τ has a bounded response
time iff workload does not exceed the processor capacity in the long term (i.e.,
$\sum_{\forall \tau_i \in \tau} U_i \leq 1$). Unfortunately, this condition does not directly generalize to
global multiprocessor scheduling with the sporadic task model introduced in
Sect. 1.1.2 (particularly, several active jobs released by the same task cannot execute
simultaneously). Some task may still have infinitely growing response time under
the condition $\sum_{\forall \tau_i \in \tau} U_i \leq M$.

The second contribution of this part is to establish the condition for the response
time of each task to be bounded in fixed-priority global scheduling. We have derived
a condition for the termination of the response time analysis procedure for an
arbitrary-deadline task τ_k, which implies the bounded response time of this task:

$$\sum_{\forall i < k} V_i^k + M \times U_k < M$$

where $V_i^k = \min(U_i, 1 - U_k)$ and $i < k$ denotes that τ_i's priority is higher than τ_k's.
Intuitively, the item $M \times U_k$ in the above inequality captures the resource waste in
the situation that all the other $M - 1$ processors are idle when the analyzed task
is executing. The refined utilization metric V_i^k restrict the interference of a higher-
priority to the part that executes not in parallel with the analyzed task τ_k.

1.3.2.2 Partitioned Scheduling

The challenge in partitioned scheduling is how to divide the system workload
to individual processors, such that the resource utilization is maximized. On
uniprocessors, Liu and Layland discovered in 1973 the utilization bound

$$\sum_{\forall \tau_i \in \tau} U_i \leq N \times (2^{\frac{1}{N}} - 1)$$

for fixed-priority uniprocessor scheduling (with the optimal RM priority assignment scheme) with implicit-deadline sporadic task systems. It has been a long-standing open problem that whether this classical result can be generalized to multiprocessor scheduling.

We close this open problem by introducing a partition-based scheduling algorithm with Liu and Layland's utilization bound in Chap. 6. The algorithm schedules workload on each processor with RM priority ordering, and guarantees success partitioning and run-time schedulability of any task set satisfying the condition

$$\sum_{\forall \tau_i \in \tau} U_i \leq M \times N \times (2^{\frac{1}{N}} - 1)$$

As mentioned in Sect. 1.1.2, the general utilization bound of any strictly partitioned scheduling algorithm is limited by 50%. To overcome this limitation, the proposed algorithm in this thesis allows a small number of tasks to migrate across different processors, and falls into the category of *partitioned scheduling with task splitting* (also called semi-partitioned scheduling).

Partitioning workload on multiprocessors shares similarities with the bin-packing problem. The bin-packing problem will become trivial if we are allowed to split the items. But this is not the case in the partitioned scheduling problem. The challenge is that the scheduling algorithm must guarantee the serial execution of a split task on different processors. The proposed algorithm uses the worst-fit bin-packing heuristics and increasing priority order for task allocation. The key insight is that the worst-fit bin-packing and increasing priority order guarantee that task splitting happens only with tasks with relatively high priorities. The high-priority tasks have larger slack times, and thereby can tolerate extra timing constraints to guarantee the serial execution of different parts of a split task on different processors.

The Liu and Layland utilization bound is tight, but may be pessimistic for a specific task set. There are a large number of task systems that exceed the Liu and Layland utilization bound but are indeed schedulable. If more information about the task system parameter is available in the design phase, it is possible to derive tighter *parametric utilization bounds* regarding available task parameter information. This thesis further extends the above algorithm to generalize most of the known parametric utilization bounds in uniprocessor real-time scheduling to multiprocessors in Chap. 7.

1.3.2.3 Cache-Aware Scheduling

Typical multi-core architectures use shared caches to boost average-case performance. The accesses to a shared cache from different cores may interfere with each other. Therefore, to precisely predict the execution time of each task, one needs to take into account the execution information of all the tasks executing simultaneously, which in turn depends on the scheduling of these tasks. This leads to a cyclic dependency between the timing behavior on the program-level and higher levels, and makes the analysis of the whole system hard or even impossible in general [2, 70].

Different from previous work [88–90] aiming at directly solving the analysis problem, Chap. 8 presents an approach that avoids the inter-core cache contention problem in the first place. We use *cache partitioning* techniques (such as page-coloring) combined with scheduling to isolate the cache spaces of hard real-time tasks running simultaneously. This yields an efficient method to control the shared cache access, in which a portion of the shared cache is assigned to each running task, and the cache replacement is restricted to each individual partition.

We assume that the shared cache is divided into partitions, and the cache space size of each task is fixed. We design cache-aware scheduling algorithms which make sure that at any time, any two running tasks' cache spaces are non-overlapped. A task can get to execute only if it gets an idle core as well as enough space (not necessarily continuous) on the shared cache. Therefore, the scheduling involves two types of resources (processing cores and caches), and the arbitration of cache resources is two dimensional (time and size). We extend the traditional analysis techniques in multiprocessor scheduling to the above described cache-aware scheduling problem, and present scheduability test conditions with compromising analysis precision and efficiency.

1.3.3   Scalable and Precise Real-Time Calculus

In this part we present techniques to improve the RTC analysis framework in both efficiency and precision. Chapter 9 deals with the high computational complexity of RTC by ruling out redundant information irrelevant to the final results from the analysis procedure. Chapter 10 improves the analysis precision of EDF scheduling in RTC.

1.3.3.1   Finitary Real-Time Calculus

The arrival/service curves in RTC are defined in the infinite range of positive real numbers. For practical implementation, RTC Toolbox [80] restricts to a class of curves that have a long-term periodicity and thus can be represented by finite data structures. In many RTC operations, the period of the output curve equals to the least common multiplier of the input periods. When many components are serially connected, the number of segments contained by the curves increases exponentially, and thus the time cost of the analysis increases exponentially as it traverses along the workload and resource flows. Due to this "period explosion" problem, RTC has an *exponential* complexity in general, and may run into serious scalability problems with complex systems.

Chapter 9 introduces a refinement of RTC, namely *Finitary Real-Time Calculus*, to solve the scalability problem. The key idea of Finitary Real-Time Calculus is to only keep a small prefix of each curve that is relevant to the final results during the whole analysis procedure. In this way, Finitary RTC drastically improves over the original RTC, and is insensitive to the parameter oddness. At the same time, Finitary RTC does not introduce any extra pessimism while getting better efficiency, i.e., it yields analysis results as precise as the original RTC.

The first crucial observation of Finitary RTC is that, the local maximal delay and backlog at each component occurs before the first intersection point of the upper arrival and lower service curves (called MBS, maximal busy-period size). Intuitively, we only need to look into the *busy periods* [32] during which the resource is continuously busy, to observe the worst-case performance. This enables us to only visit the input curves up to MBS at each component to precisely obtain its maximal backlog and delay.

The second crucial observation is that at each component we can use input finitary curves to generate finitary curves at output. This is non-trivial, since the min-plus and max-plus deconvolution operations

$$(f \oslash g)(\Delta) = \sup_{\lambda \geq 0} \{f(\Delta + \lambda) - g(\lambda)\}$$

$$(f \overline{\oslash} g)(\Delta) = \inf_{\lambda \geq 0} \{f(\Delta + \lambda) - g(\lambda)\}$$

require to check the value of $f(\Delta + \lambda) - g(\lambda)$ for *all* $\lambda \geq 0$. To solve this problem, we use a finitary version of the deconvolution operations in the computation of the output arrival and service curves. We prove that to calculate the output curve up to interval size x, it is enough to only visit the part of input curves up to interval size $x +$ MBS.

However, the MBS value for each component is not revealed until its input arrival and service curves are actually known. To address this problem, we use safe approximations of the input curves to quickly pre-analyze the whole network and obtain safe estimation of the MBS value for each component. Then we can backtrack the analysis network with MBS estimations to decide the size of the input finitary curves for each component, and eventually decide the size of the curves we need to keep at the initial inputs.

The computational complexity of Finitary RTC is pseudo-polynomial, in contrast to the exponential complexity of the original RTC. Experimental evaluations with case studies and randomly generated systems show that Finitary RTC can drastically improve the analysis efficiency, especially for systems with complex timing characteristics.

1.3.3.2 Analysis of EDF in Real-Time Calculus

The RTC theory was originally established with a specific type of components GPC [74], which naturally models the FPS policy for resource arbitration among different workload streams. EDF is another important scheduling policy in real-time systems.

Recently, RTC has also been extended to model EDF scheduling, but with rather imprecise analysis backbone. The current method [91] to compute output arrival curves of an EDF component consists of two steps. It first verifies whether the computational request of all streams can be served by its deadline. If yes, the output curves are generated assuming that the processing of each request completes just before the deadline. This is clearly over-pessimistic, since in many cases the requests may be served much earlier than their deadlines even in the worst-case, especially when the total workload is much lower than the resource service.

Apart from the imprecise computation of output curves, the above method also limits the modeling power of RTC. The deadline of each request in EDF can be viewed as the metric to decide its priority, i.e., the so-called *priority point* [92, 93]. The priority point is not necessarily the same as the deadline, but can be any instant in time. For example, if the priority point of each request aligns with its invocation time, the requests will be scheduled in a FIFO manner. However, the current RTC framework interprets EDF scheduling in a narrow sense and in general does not allow to model EDF-like scheduling algorithms with priority points different from deadlines.

Chapter 10 presents a new method to analyze EDF components in RTC. The key is to use worst-case response times, instead of deadlines, in the calculation of output arrival curves. This method not only improves the analysis precision, but also decouples the concept of deadline and priority point in EDF scheduling, and thereby supports modeling and analysis of a wide range of EDF-like scheduling policies in RTC.

The technical challenge is how to precisely bound the response time of each request under EDF scheduling in RTC. Response time analysis techniques for EDF have been developed for restricted task models [94], based on the explicitly enumeration of a (potentially very large) number of release patterns that may lead to the worst-case response time. This is, on one hand, computationally expensive, and on the other hand difficult to handle the abstract workload and resource model in RTC.

In Chap. 10, we develop a new response time analysis technique for EDF scheduling, which upper bounds the response time *indirectly* by calculating a lower bound of the slack time. We first present a simple but over-approximate analysis method which lower-bounds the slack time by measuring the maximal horizontal distance between the lower service curve and the curve of *demand bound function* [30] (derived from the upper arrival curve). Based on the insights of the first analysis, we then develop an exact response time analysis method at the cost of increased computational complexity. Finally, we use a case study to demonstrate the precision improvement by our techniques for the analysis of EDF components in RTC.

Part I
Cache Analysis for WCET Estimation

Chapter 2
MRU Cache Analysis for WCET Estimation

Most previous work on cache analysis for WCET estimation assumes a particular replacement policy LRU. In contrast, much less work has been done for non-LRU policies, since they are generally considered to be very unpredictable. However, most commercial processors are actually equipped with these non-LRU policies, since they are more efficient in terms of hardware cost, power consumption, and thermal output, while still maintaining almost as good average-case performance as LRU.

In this chapter, we study the analysis of MRU, a non-LRU replacement policy employed in mainstream processor architectures like Intel Nehalem. Our work shows that the predictability of MRU has been significantly under-estimated before, mainly because the existing cache analysis techniques and metrics do not match MRU well. As our main technical contribution, we propose a new cache hit/miss classification, k-Miss, to better capture the MRU behavior, and develop formal conditions and efficient techniques to decide k-Miss memory accesses. A remarkable feature of our analysis is that the k-Miss classifications under MRU are derived by the analysis result of the same program under LRU. Therefore, our approach inherits the advantages in efficiency and precision of the state-of-the-art LRU analysis techniques based on abstract interpretation. Experiments with instruction caches show that our proposed MRU analysis has both good precision and high efficiency, and the obtained estimated WCET is rather close to (typically 1–8 % more than) that obtained by the state-of-the-art LRU analysis, which indicates that MRU is also a good candidate for cache replacement policies in real-time systems.

© Springer International Publishing Switzerland 2016
N. Guan, *Techniques for Building Timing-Predictable Embedded Systems*,
DOI 10.1007/978-3-319-27198-9_2

2.1 Introduction

For hard real-time systems one must ensure that all timing constraints are satisfied. To provide such guarantees, a key problem is to bound the worst-case execution time (WCET) of programs [4]. To derive safe and tight WCET bounds, the analysis must take into account the timing effects of various micro-architecture features of the target hardware platform. Cache is one of the most important hardware components affecting the timing behavior of programs: the timing delay of a cache miss could be several orders of magnitude greater than that of a cache hit. Therefore, analyzing the cache access behavior is a key problem in WCET estimation. However, the cache analysis problem of statically determining whether each memory access is a hit or a miss is challenging.

Much work has been done on cache analysis for WCET estimation in the last two decades. Most of the published works assume a particular cache replacement policy, called LRU (Least-Recently-Used), for which researchers have developed successful analysis techniques to precisely and efficiently predict cache hits/misses [4]. In contrast, much less attention has been paid to other replacement policies like MRU (Most-Recently-Used)[1] [97], FIFO (First-In-First-Out) [21], and PLRU (Pseudo-LRU) [96]. In general, research in the field of real-time systems assumes LRU as the default cache replacement policy. Non-LRU policies in general, in fact, are considered to be much less predictable than LRU, and it would be very difficult to develop precise and efficient analyses for them. It is recommended to only use LRU caches when timing predictability is a major concern in the system design [27].

However, most commercial processors actually do not employ the LRU cache replacement policy. The reason is that the hardware implementation logic of LRU is rather expensive [81], which results in higher hardware cost, power consumption, and thermal output. On the other hand, non-LRU replacement policies like MRU, FIFO, and PLRU enjoy simpler implementation logic, but still have almost as good average-case performance as LRU [82]. Therefore, hardware manufacturers tend to choose these non-LRU replacement policies in processor design, especially for embedded systems subject to strict cost, power, and thermal constraints.

In this chapter, we study one of the most widely used cache replacement policies MRU. MRU uses a mechanism similar to the clock replacement algorithm in virtual memory mapping [98]. It only uses one bit for each cache line to maintain age information, which is very efficient in hardware implementation. MRU has been employed in mainstream processor architectures like Intel Nehalem (the architecture codename of processors like Intel Xeon, Core i5, and i7) [99] and UltraSPARC T2 [100]. A previous work comparing the average-case performance

---

[1] The name of the MRU replacement policy is inconsistent in the literature. Sometimes, this policy is called Pseudo-LRU because it can be seen as a kind of approximation of LRU. However, we use the name MRU to keep consistency with previous works in WCET research [26, 95], and to distinguish it from another Pseudo-LRU policy PLRU [96] which uses tree structures to store access history information.

of cache replacement policies with the SPEC CPU2000 benchmark showed that MRU has almost as good average-case performance as LRU [82]. To the best of our knowledge, there has been no previous work dedicated to the analysis of MRU in the context of WCET estimation. The only relevant work was performed by Reineke et al. [26] and Reineke and Grund [101], which studies general timing predictability properties of different cache replacement policies. The cited work argues that MRU is a very unpredictable policy.

However, this chapter shows that the predictability of MRU actually has been significantly under-estimated. The state-of-the-art cache analysis techniques are based on *qualitative* classifications, to determine whether the memory accesses related to a particular point in the program are always hits or not (except the first access that may be a cold miss). This approach is highly effective for LRU since most memory accesses indeed exhibit such a "black or white" behavior under LRU. In this work we show that the memory accesses may have more nuanced behavior under MRU: a small number of the accesses are misses while all the other accesses are hits. By the existing analysis framework based on qualitative classifications, such a behavior has to be treated as if all the accesses are misses, which inherently leads to very pessimistic analysis results.

In this chapter, we introduce a new cache hit/miss classification k-Miss (at most k accesses are misses while all the others are hits). In contrast to qualitative classifications, k-Miss can *quantitatively* bound the number of misses incurred at certain program points, hence it can more precisely capture the nuanced behavior in MRU. As our main technical contribution, we establish formal conditions to determine k-Miss memory accesses, and develop techniques to efficiently check these conditions. Notably, our technique uses the cache analysis results of the same program under LRU to derive k-Miss classification under MRU. Therefore, our technique inherits the advantages in both efficiency and precision from the state-of-the-art LRU analysis based on *abstract interpretation* (AI) [19].

We conduct experiments with benchmark programs with *instruction* caches to evaluate the quality of our proposed analysis, which show that our MRU analysis has both good precision and efficiency: the estimated WCET obtained by our MRU analysis is on average 2–10 % more than that obtained by simulations, and the analysis of each benchmark program terminates within 0.1 s on average. Moreover, the estimated WCET by our MRU analysis is close to (typically 1–8 % more than) that obtained by the state-of-the-art LRU analysis. This suggests that MRU is also a good candidate for instruction cache replacement policies in real-time systems, especially considering MRU's other advantages in hardware cost, power consumption, and thermal output.

Although the experimental evaluation in this chapter only considers instruction caches, the properties of MRU disclosed in this chapter also hold for data caches and our analysis techniques can be directly applied to systems with data caches. We didn't include experiments with data caches because predicting data cache behaviors heavily relies on value analysis [4], which is another important topic in WCET estimation but orthogonal to the cache analysis issue studied in this chapter.

Since our prototype does not yet support high-quality value analysis functionalities, we currently cannot provide a meaningful evaluation with data caches. The evaluation of the proposed MRU analysis with data caches is left as our future work.

2.2 Related Work

Most previous work on cache analysis for static WCET estimation assumes the LRU replacement policy. Li and Malik [18] and Li et al. [102] use integer linear programming (ILP)-only approaches where the cache behavior prediction is formulated as part of the overall ILP problem. These approaches suffer from serious scalability problems due to the exponential complexity of ILP, and thus cannot handle realistic programs on modern processors. Arnold et al. [103] and Mueller [104, 105] proposed a technique called *static cache simulation*, which iteratively calculates the instructions that *may* be in the cache at the entry and exit of each basic block until the collective cache state reaches a fixed point, and then uses this information to categorize the caching behavior of each instruction.

A milestone in the research of static WCET estimation is establishing the framework combining micro-architecture analysis by *abstract interpretation* (AI) and path analysis by *implicit path enumeration technique* (IPET) [19]. The AI-based cache analysis statically categorizes the caching behavior of each instruction by sound Must, May, and Persistence analyses, which have both high efficiency and good precision for LRU caches. The IPET-based path analysis uses the cache behavior classification to derive a delay invariant for each instruction and encodes the WCET calculation problem into ILP formulation. Such a framework forms the common foundation for later research in cache analysis for WCET estimation. For example, it has been refined and extended to deal with nested loops [106, 107], data caches [108–110], multi-level caches [111, 112], shared caches [113, 114], and cache-related preemption delay [115, 116].

In contrast, much less work has been done for non-LRU caches. Although some important progress has been made in the analysis of policies like FIFO [21, 28] and PLRU [23], in general these analyses are much less precise than for LRU. To the best of our knowledge, there has been no work dedicated to the analysis of MRU in the context of WCET estimation.

Reineke et al. [26], Reineke and Grund [101, 117] and Reineke [95] have conducted a series of fundamental studies on predictability properties of different cache replacement policies. Reineke et al. [26] defines several *predictability* metrics, regarding the minimal number of different memory blocks that are needed to (a) completely clear the original cache content (evict), (b) reach a completely known cache state (fill), (c) evict a block that has just been accessed (mls). Reineke and Grund [117] studies the *sensitivity* of different cache replacement policies, which expresses to what extent the initial state of the cache may influence the number of cache hits and misses during program execution. According to all the above metrics, LRU appears significantly more predictable than other policies like MRU, FIFO, and PLRU. Reineke and Grund [101] studies the *relative competitiveness* between different policies by providing upper (lower) bounds of the ratio on the number of misses (hits) between two different replacement policies during the whole program execution. By such information, one can use the cache analysis result under one replacement policy to predict the number of cache misses (hits) of the program under another policy. This approach is different in many ways from our proposed analysis based on k-Miss classification. Firstly, while the relative competitiveness approach provides bounds on the number of misses of the *whole program*,[2] the k-Miss classification bounds the number of misses at individual program points. Secondly, while the bounds on the number of misses provided by the relative competitiveness analysis are linear with respect to the total number of accesses, our k-Miss analysis provides constant bounds. Thirdly, the k-Miss classification for MRU does not necessarily rely on the analysis result of LRU, and one can identify k-Miss by other means, e.g., directly computing the maximal stack distance as defined in Sect. 2.4. Overall, our proposed analysis based on k-Miss can better capture MRU cache behavior and support a much more precise WCET estimation than the relative competitiveness approach.

Finally, we refer to [4, 118] for comprehensive surveys on WCET analysis techniques and tools, which cover many relevant references that are not listed here.

2.3   Basic Concepts

We assume an abstract processor architecture model: The processor has a perfect pipeline and instructions are fetched sequentially. The processor has a cache between the processing core and the main memory. The execution delay of each instruction only depends on whether the corresponding memory content is in the cache or not, and the time to deliver data from the main memory to the cache is constant. Other factors affecting the execution delay are not considered.

We assume that the cache is *set-associative* or *fully-associative*. In set-associative caches, the accesses to memory references mapped to different cache sets do not affect each other, and each cache set can be treated as a fully-associative cache and analyzed independently. We present the cache analysis techniques in the context of a fully-associative cache for simplicity of presentation, and the experiments are all conducted with *set-associative* caches. Let the cache have L ways, i.e., the cache consists of L *cache lines*. The memory content that fits into one cache line is called a *memory block*.

We consider the common class of programs represented by control-flow graphs (CFG). Programs that are difficult to be modeled by CFGs, e.g., self-modified

---

[2]The relative competitiveness can also be used as Must/May analysis to predict the cache access behavior at individual program points. However, this relies on the analysis under other policies with typically a much smaller cache sizes (to get 1-competitiveness), which generally yields very pessimistic results.

programs, are usually not suitable for safe-critical systems and out of our scope. A CFG can be defined on the basis of individual nodes as follows:

Definition 2.1 (CFG on the Basis of Nodes). A CFG is a tuple $G = (N, E, n_{st})$:

- $N = \{n_1, n_2, \cdots\}$ is the set of *nodes* in the CFG;
- $E = \{e_1, e_2, \cdots\}$ is the set of directed *edges* in the CFG;
- $n_{st} \in N$ is the unique *starting node* of the CFG.

A CFG can also be represented as a digraph of *basic blocks* [119]:

Definition 2.2 (CFG on the Basis of Basic Blocks). A CFG is a tuple $G = (B, E, b_{st})$:

- $B = \{b_1, b_2, \cdots\}$ is the set of *basic blocks* in the CFG;
- $E = \{e_1, e_2, \cdots\}$ is the set of directed *edges* connecting the basic blocks in the CFG;
- $b_{st} \in B$ is the unique *starting basic block* of the CFG.

Figure 2.1 shows a CFG example on the basis of individual nodes and basic blocks respectively. Letter $a, b, \cdots$ inside each node denotes the memory block accessed by the node. When we mention the CFG in the rest of this chapter, it is by default on the basis of nodes unless otherwise specified.

At run-time, when (a node of) the program accesses a memory block, the processor first checks whether the memory block is in the cache. If yes, it is a *hit*, and the program directly accesses this memory block from the cache. Otherwise, it is a *miss*, and this memory block is first installed in the cache before the program accesses it.

A memory block only occupies one cache line regardless of how many times it is accessed. So the number of *unique* accesses to memory blocks, i.e., the number of *pairwise different* memory blocks in an access sequence is important to the cache behavior. We use the following concept to reflect this:

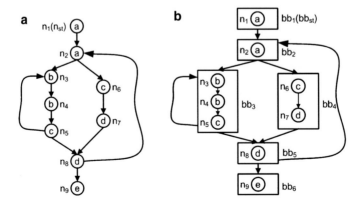

Fig. 2.1 A control-flow-graph example. (**a**) On the basis of nodes, (**b**) on the basis of basic blocks

Definition 2.3 (Stack Length). The Stack Length of a memory access sequence corresponding to a path p in the CFG, denoted by $\pi(p)$, is the number of pairwise different memory blocks accessed along p.

For example, the stack length of the access sequence

$$a \rightarrow b \rightarrow c \rightarrow a \rightarrow d \rightarrow a \rightarrow b \rightarrow d$$

is 4, since only 4 memory blocks a, b, c, and d are accessed in this sequence.

The number of memory blocks accessed by a program is typically far greater than the number of cache lines, so a replacement policy must decide which block to be replaced upon a miss. In the following we describe the LRU and MRU replacement policy, respectively.

2.3.1 LRU *Replacement*

The LRU replacement policy always stores the most recently accessed memory block in the first cache line. When the program accesses a memory block s, if s is not in the cache (miss), then all the memory blocks in the cache will be shifted one position to the next cache line (the memory block in the last cache line is removed from the cache), and s is installed to the first cache line. If s is in the cache already (hit), then s is moved to the first cache line and all memory blocks that were stored before s's old position will be shifted one position to the next cache line. Figure 2.2 illustrates the update upon an access to memory block s in an LRU cache of 4 lines. In the figure, the uppermost block represents the first (lowest-index) cache line and the lowermost block is the last (highest-index) one. All figures in this chapter follow this convention.

A metric defined in [26] to evaluate the predictability of a replacement policy is the *minimal-life-span* (mls), the minimal number of pairwise different memory blocks required to evict a just visited memory block out of the cache (not counting the access that brought the just visited memory block into the cache). It is known that [26]:

Lemma 2.1. *The* mls *of* LRU *is L.*

Recall that L is the number of lines in the cache. The mls metric can be directly used to determine cache hits/misses for a memory access sequence: if the stack length of

Fig. 2.2 Illustration of LRU cache update with $L = 4$, where the left part is a miss and the right part is a hit

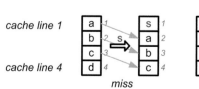

the sequence between two successive accesses to the same memory block is smaller than mls, then the later access must be a hit. For example, for a memory access sequence

$$a \rightarrow b \rightarrow c \rightarrow c \rightarrow d \rightarrow a \rightarrow e \rightarrow b$$

on a 4-way LRU cache, we can easily conclude that the second access to memory block a is a hit since the sequence between two accesses to a is $b \rightarrow c \rightarrow c \rightarrow d$, which has stack length 3. The second access to b is a miss since the stack length of the sequence $c \rightarrow c \rightarrow d \rightarrow a \rightarrow e$ is 4. Clearly, replacement policies with larger mls are preferable, and the upper bound of mls is L.

2.3.2 MRU *Replacement*

For each cache line, the MRU replacement policy stores an extra MRU-bit, to approximately represent whether this cache line was recently visited. An MRU-bit at 1 indicates that this line was recently visited, while at 0 indicates the opposite. Whenever a cache line is visited, its MRU-bit will be set to 1. Eventually there will be only one MRU-bit at 0 in the cache. When the cache line with the last MRU-bit at 0 is visited, this MRU-bit is set to 1 and all the other MRU-bits change back from 1 to 0, which is called a *global-flip*.

More precisely, when the program accesses a memory block s, MRU replacement first checks whether s is already in the cache. If yes, then s will still be stored in the same cache line and its MRU-bit is set to 1 regardless of its original state. If s is not in the cache, MRU replacement will find the first cache line whose MRU-bit is 0, then replace the originally stored memory block in it by s and set its MRU-bit to 1. After the above operations, if there still exists some MRU-bit at 0, the remaining cache lines' states are kept unchanged. Otherwise, all the remaining cache lines' MRU-bits are changed from 1 to 0, which is a global-flip. Note that the global-flip operation guarantees that at any time there is at least one MRU-bit in the cache being 0.

In the following we present the MRU replacement policy formally. Let M be the set of all the memory blocks accessed by the program plus an element representing emptiness. The MRU cache state can be represented by a function $C : \{1, \cdots, L\} \rightarrow M \times \{0, 1\}$. We use $C(i)$ to denote the state of the ith cache line. For example, $C(i) = (s, 0)$ represents that cache line i currently stores memory block s and its MRU-bit is 0. Further, we use $C(i).\omega$ and $C(i).\beta$ to denote the resident memory block and the MRU-bit of cache line i. The update rule of MRU replacement can be described by the following steps, where C and C' represent the cache state before and after the update upon an access to memory block s, respectively, and δ denotes the cache line where s should be stored after the access:

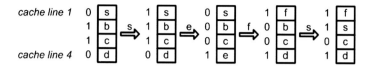

Fig. 2.3 An example illustrating MRU and its mls

1. If there exists h s.t. $C(h).\omega = s$, then let $\delta \leftarrow h$, otherwise let $\delta = h$ s.t. $C(h).\beta = 0$ and $C(j).\beta = 1$ for all $j < h$.
2. $C'(\delta) \leftarrow (s, 1)$
3. If $C(h).\beta = 1$ for all $h \neq \delta$, then let $C'(j) \leftarrow (C(j).\omega, 0)$ for all $j \neq \delta$ (i.e., global-flip), otherwise $C'(j) \leftarrow C(j)$ for all $j \neq \delta$.

Figure 2.3 illustrates MRU replacement with a 4-way cache. First the program accesses memory block s, which is already in the cache. So s still stays in the same cache line, and the corresponding MRU-bit is changed to 1. Then the program accesses e, which is not in the cache yet. Since only the 4th cache line's MRU-bit is 0, e is installed in that line and triggers the global-flip, after which the 4th cache line's MRU-bit is 1 and all the other MRU-bits are changed to 0. Then the program accesses f and s in order, which are both not in the cache, so they will be installed to the first and second cache line with MRU-bits at 0 and change these bits to 1.

In MRU caches, an MRU-bit can roughly represent how old the corresponding memory block is, and the replacement always tries to evict a memory block that is relatively old. So MRU can be seen as an approximation of LRU. However, such an approximation results in a very different mls [26]:

Lemma 2.2. *The* mls *of* MRU *is* 2.

The example in Fig. 2.3 illustrates this lemma, where only two memory blocks e and f are enough to evict a just-visited memory block s. It is easy to extend this example to arbitrarily many cache lines, where we still only need two memory blocks to evict s. Partly due to this property, MRU has been believed to be a very unpredictable replacement policy, and to the best of our knowledge it has never been seriously considered as a good candidate for timing-predictable architectures.

2.4 A Review of the Analysis for LRU

As we mentioned in Sect. 2.1, the MRU analysis proposed in this chapter uses directly the results of the LRU analysis for the same program. Thus, before presenting our new analysis technique, we first provide a brief review of the state-of-the-art analysis technique for LRU.

Exact cache analysis suffers from a serious state-space explosion problem. Hence, researchers resort to approximation techniques separating path analysis and cache analysis for good scalability [19]. Path analysis requires an upper bound on the timing delay of a node whenever it is executed. Therefore, the main purpose of the LRU cache analysis is to decide the cache hit/miss classification (CHMC) for each node [19, 103]:

- AH (always hit): The node's memory access is always hit whenever it is executed.
- FM (first miss): The node's memory access is miss for the first execution, but always hit afterwards. This classification is useful to handle "cold miss" in loops.
- AM (always miss): The node's memory access is always miss whenever it is executed.
- NC (non-classified): Cannot be classified into any of the above categories. This category has to be treated as AM in the path analysis.

Among the above CHMC, we call AH and FM *positive classification* since they ensure that (the major portion of) the memory accesses of a node to be hits, and call AM and NC *negative classification*.

Recall that the mls of LRU is L, and one can directly use this property to decide the hit/miss of a node with linear access sequences. However, a CFG is generally a digraph, and there may be multiple paths between two nodes.

The following concept captures the maximal number of pairwise different memory blocks between two nodes accessing the same memory block in the CFG.

Definition 2.4 (Maximal Stack Distance). Let n_i and n_j be two nodes accessing the same memory block s. The Maximal Stack Distance from n_i to n_j, denoted by $dist(n_i, n_j)$, is defined as:

$$dist(n_i, n_j) = \begin{cases} \max\{\pi(p) \mid p \in P(n_i, n_j)\} & \text{if } P(n_i, n_j) \neq \emptyset \\ 0 & \text{if } P(n_i, n_j) = \emptyset \end{cases}$$

where $P(n_i, n_j)$ is the set of paths satisfying

- n_i and n_j is the first and last node of the path, respectively;
- None of the nodes in the path, except the first and last, accesses s.

Note that the maximal stack distance between two nodes is direction sensitive, i.e., $dist(n_i, n_j)$ may not be equal to $dist(n_j, n_i)$. The example in Fig. 2.4 illustrates the maximal stack distance using a CFG with three nodes n_1, n_3, and n_7 accessing the same memory block s. We have $dist(n_1, n_7) = 5$ since $P(n_1, n_7)$ contains a path

$$n_1 \rightarrow n_4 \rightarrow n_5 \rightarrow n_8 \rightarrow n_4 \rightarrow n_6 \rightarrow n_8 \rightarrow n_4 \rightarrow n_7$$

in which s, a, c, d, and e are accessed. We have $dist(n_1, n_3) = 2$ since $n_1 \rightarrow n_2 \rightarrow n_3$ is the only path in $P(n_1, n_3)$ (any other path from n_1 to n_3 does not satisfy the second condition for P). We have $dist(n_3, n_7) = 0$ since any path from n_3 to n_7 has to go through n_1 which also accesses s.

Now one can use the maximal stack distance to judge whether the CHMC of a node n_i is positive: n_j falls into the positive classification (AH or FM), if $dist(n_i, n_j) \leq L$ holds for any node n_i that accesses the same memory block s as n_j.

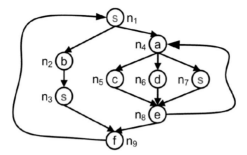

Fig. 2.4 Illustration of Maximal Stack Distance

This is because there are not enough pairwise different memory blocks to evict s along any path to n_i since the last access to s.

However, computing the exact maximal stack distance is in general very expensive. Therefore, the LRU analysis resorts to over-approximation by abstract interpretation. The main idea is to define an abstract cache state and iteratively traverse the program until the abstract state converges to a fixed point, and use the abstract state of this fixed point to determine the CHMC. There are mainly three fixed-point analyses:

- **Must** analysis to determine AH nodes.
- **May** analysis to determine AM nodes.
- **Persistence** analysis to determine FM nodes.

A node is an NC if it cannot be classified by any of the above analyses. We refer to [24, 110] for details of these fixed-point analyses.

2.5 The New Analysis of MRU

In this section we present our new analysis for MRU. First we show that the existing CHMC in the LRU analysis as introduced in last section is actually not suitable to capture the cache behavior under MRU, and thus we introduce a new classification k-Miss (Sect. 2.5.1). After that we introduce the conditions for nodes to be k-Miss (Sect. 2.5.2), and show how to efficiently check these conditions (Sect. 2.5.3). Then the k-Miss classification is generalized to more precisely analyze nested-loops (Sect. 2.5.4). Finally we present how to apply the cache analysis results in the path analysis to obtain the WCET estimation (Sect. 2.5.5).

2.5.1 New Classification: k-Miss

First we consider the example in Fig. 2.5a. We can see that $\text{dist}(n_1, n_1) = 4$, i.e., 4 pairwise different memory blocks appear in each iteration of the loop no matter

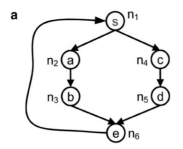

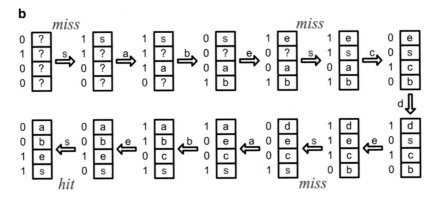

Fig. 2.5 An example motivating the *k*-Miss classification. (**a**) A CFG example, (**b**) cache update when the two branches are taken alternatively

which branch is taken. Since $\mathsf{dist}(n_1, n_1)$ is larger than 2 (the mls of MRU), n_1 cannot be decided as a positive classification using mls.

Now we have a closer look into this example, considering a particular execution sequence in which the two branches are taken alternatively, as shown in Fig. 2.5b. Assume that the memory blocks initially stored in the cache (denoted by "?") are all different from the ones that appear in Fig. 2.5a, and initial MRU-bits are shown in the first cache state of Fig. 2.5b.

We can see that the first three executions of s are all misses. The first miss is a cold miss which is unavoidable anyway under our initial cache state assumption. However, the second and third accesses are both misses because s is evicted by other memory blocks. Indeed, node n_1 cannot be determined as AH or FM, and one has to put it into the negative classification and treat it as being always miss whenever it is executed.

However, if the sequence continues, we can see that when n_1 is visited for the fourth time, s is actually in the cache, and most importantly, *the access of n_1 will* *always be a hit afterwards* (we do not show a complete picture of this sequence, but this can be easily seen by simulating the update for a long enough sequence until a cycle appears).

The existing positive classification AH and FM is inadequate to capture the behavior of nodes like n_1 in the above example, which only encounters a smaller number of misses, but will eventually go into a stable state of being always hits. Such behavior is actually quite common under MRU. Therefore, the analysis of MRU will be inherently very pessimistic if one only relies on the AH and FM classification to claim cache hits.

The above phenomenon shows the need for a more precise classification to capture the MRU cache behavior. As we show in Sect. 2.5.2, the number of misses under MRU may be bound not only for individual nodes, but also for a set of nodes that access the same memory block. This leads us to the definition of the k-Miss classification as follows:

Definition 2.5 (k-Miss). A set of nodes $S = \{n_1, \cdots, n_i\}$ is k-Miss iff at most k accesses by nodes in S are misses while all the other accesses are hits.

The traditional classification FM can be viewed as a special case of k-Miss with a singleton node set and $k = 1$. Note that although the k-Miss classification can bound the number of misses for a set of nodes, it does not say anything about when do these k times of misses actually occur. The misses do not necessarily occur at the first k accesses of these nodes. It allows the misses and hits to appear alternatively, as long as the total number of misses does not exceed k.

2.5.2 Conditions for k-Miss

In this section we establish the conditions for a set of nodes to be k-Miss. We start with an important property of MRU:

Lemma 2.3. *At least k pairwise different memory blocks are needed to evict a memory block in cache line k with MRU-bit at 1.*

Proof. Only the memory block in a cache line with MRU-bit at 0 can be evicted, so before the eviction of s there must be a global-flip to change the MRU-bit of cache line k from 1 to 0. Right after the global flip, the number of 0-MRU-bits among cache lines $\{1, \cdots, k\}$ is at least $k - 1$, so $k - 1$ pairwise different memory blocks (which are also different from the one triggering the global-flip) are needed to fill up these 0-MRU-bit cache lines. In total, the number of pairwise different memory blocks required is at least k. □

Lemma 2.3 indicates that the minimal-life-span of memory blocks installed to different cache lines are asymmetric: a cache line with a greater index provides a larger minimal-life-span guarantee (while the mls metric does not distinguish different positions but simply captures the worst case). To provide a better analysis than the mls approach, one needs information about where a memory block is installed. However, under MRU a memory block may be installed to any cache line without restricting the cache state beforehand. Since the initial cache state is unknown, and the precise cache state information is lost quickly during the abstract analysis, it is difficult to precisely predict the position of a memory block in the cache.

However, Lemma 2.3 indeed gives us opportunities to do a better analysis. When a memory block is installed to a cache line with a larger index, it becomes more difficult to be evicted. So the main idea of our analysis is to verify whether a memory block will eventually be installed to a "safe position" (a cache line with large enough index) and stay there afterwards (as long as it executes in the scope of the program under analysis). The k times of misses in k-Miss happens before the memory block is installed to the "safe position," and after that all the accesses will be hits. In the following we show the condition for a memory block to have such behavior. We first introduce an auxiliary lemma:

Lemma 2.4. *On an L-way* MRU *cache, L pairwise different memory blocks are accessed between two successive global-flips (including the ones triggering these two global-flips).*

Proof. Right after a global-flip, there are $L - 1$ cache lines whose MRU-bits are 0. In order to have the next flip, all these cache lines of which the MRU-bits are 0 need to be accessed, i.e., it needs $L - 1$ pairwise different memory blocks that are also different from the one causing the first global-flip. So in total L pairwise different memory blocks are involved in the access sequence between two successive global-flips.

Lemma 2.4 is illustrated by the example in Fig. 2.6 with $L = 4$. The access to memory block a triggers the first global-flip, after which 3 MRU-bits are 0. To trigger the next global-flip, these three MRU-bits have to be changed to 1, which needs 3 pairwise different memory blocks. So in total 4 pairwise different memory blocks are involved in the access sequence between these two global-flips. With this auxiliary lemma, we are able to prove the following key property:

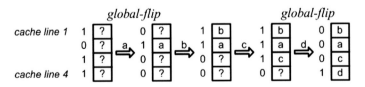

Fig. 2.6 Illustration of Lemma 2.4

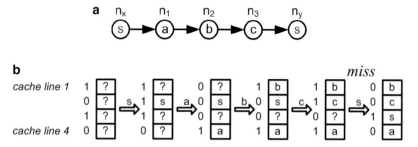

Fig. 2.7 Illustration of Lemma 2.5. (**a**) A path from n_x to n_y, (**b**) s is moved to a larger index when it is loaded back

Lemma 2.5. *Suppose that under* MRU *at some point a memory block s is accessed by node n_x at cache line i (either hit or miss), and the next access to s is a miss caused by n_y upon which s is installed to cache line j. We have $j > i$ if the following condition holds:*

$$\mathrm{dist}(n_x, n_y) \leq L. \tag{2.1}$$

Figure 2.7 illustrates Lemma 2.5, where n_x and n_y are two nodes accessing the same memory block s and satisfying Condition (2.1). We focus on a particular path as shown in Fig. 2.7a. Figure 2.7b shows the cache update along this path: first n_x accesses s in the second cache line. After s is evicted out of the cache and is loaded back again, it is installed to the third cache line, which is one position below the previous one. In the following we give a formal proof of the lemma.

Proof. Let event ev_x be the access to s at cache line i by n_x as stated in the lemma, and event ev_y the installation of s to cache line j by n_y. We prove the lemma by contradiction, assuming $j \leq i$.

The first step is to prove that there are at least two global-flips in the event sequence $\{ev_{x+1}, \cdots, ev_{y-1}\}$ (ev_{x+1} denotes the event right after ev_x and ev_{y-1} the event right before ev_y).

Before ev_y, s has to be first evicted out of the cache. Let event ev_v denote such an eviction of s, which occurs at cache line i. By the MRU replacement rule, a memory block can be evicted from the cache only if the MRU-bit of its resident cache line is 0. So we know $C(i).\beta = 0$ *right before* ev_v.

On the other hand, we also know that $C(i).\beta = 1$ *right after* event ev_x. And since only a global-flip can change an MRU-bit from 1 to 0, we know that there must exist at least one global-flip among the events $\{ev_{x+1}, \cdots, ev_{v-1}\}$.

Then we focus on the event sequence $\{ev_v, \cdots, ev_{y-1}\}$. We distinguish two cases:

- $i = j$. Right after the eviction of s at cache line i (event ev_v), the MRU-bit of cache line i is 1. On the other hand, just before the installation of s to cache line j (event ev_y), the MRU-bit of cache line j must be 0. Since $i = j$, there must be at

least one global-flip among the events $\{ev_{v+1}, \cdots, ev_{y-1}\}$, in order to change the
MRU-bit of cache line $i = j$ from 1 to 0.

- $i > j$. By the MRU replacement rule, we know that just before s is evicted in
  event ev_v, it must be true that $\forall h < i : C(h).\beta = 1$, and hence $C(j).\beta = 1$.
  On the other hand, just before the installation of s in event ev_y, the MRU-bit of
  cache line j must be 0. Therefore, there must be at least one global-flip among
  the events $\{ev_v, \cdots, ev_{y-1}\}$, in order to change the MRU-bit of cache line j from
  1 to 0.

In summary, there is at least one global-flip among $\{ev_v, \cdots, ev_{y-1}\}$.

Therefore, we can conclude that there are at least two global-flips among the
events $\{ev_{x+1}, \cdots, ev_{y-1}\}$. By Lemma 2.4 we know that at least L pairwise different
memory blocks are accessed in $\{ev_{x+1}, \cdots, ev_{y-1}\}$. Since ev_y is the first access to
memory block s after ev_x, there is no access to s in $\{ev_{x+1}, \cdots, ev_{y-1}\}$, so at least
$L + 1$ pairwise different memory blocks are accessed in $\{ev_x, \cdots, ev_y\}$.

On the other hand, let p be the path that leads to the sequence $\{ev_x, \cdots, ev_y\}$.
Clearly, p starts with n_x and ends with n_y. We also know that no other node along
p, apart from n_x and n_y, accesses s, since ev_y is the first event accessing s after
ev_x. So p is a path in $P(n_x, n_y)$ (Definition 2.4), and we know $\mathsf{dist}(n_x, n_y) \geq \pi(p)$.
Combining this with Condition (2.1) we have $\pi(p) \leq L$, which contradicts with that
at least $L+1$ pairwise different memory blocks are accessed in $\{ev_x, \cdots, ev_y\}$ as we
concluded above.

To see the usefulness of Lemma 2.5, we consider a special case where only
one node n in the CFG accesses memory block s and $\mathsf{dist}(n, n) \leq L$ as shown in
Fig. 2.8a. In this case, by Lemma 2.5 we know that each time s is accessed (except
the first time), there are only two possibilities:

- the access to s is a hit, or
- the access to s is a miss and s is installed to a cache line with a strictly larger
  index than before.

So we can conclude that the access to s can only be miss for at most L times since
the position of s can only "move downwards" for a limited number of times which
is bounded by the number of cache lines. Moreover, we can combine Lemma 2.3
and Lemma 2.5 to have a stronger claim: if condition $\mathsf{dist}(n, n) \leq k$ holds for some
$k \leq L$, then the access to s can only be miss for at most k times, since the number of
pairwise different memory blocks along the path from n back to n is not enough to
evict s as soon as it is installed to cache line k.

However, in general there could be more than one node in the CFG accessing
the same memory block, where Lemma 2.5 cannot be directly applied to determine
the k-Miss classification. Consider the example in Fig. 2.8b, where two nodes n_1
and n_2 both access the same memory block s, and we have $\mathsf{dist}(n_1, n_2) \leq L$
and $\mathsf{dist}(n_2, n_1) > L$. In this case, we cannot classify n_2 as a k-Miss, although
Lemma 2.5 still applies to the path from n_1 to n_2. This is because Lemma 2.5 only
guarantees the position of s will move to larger indices each time n_2 encounters a

Fig. 2.8 An example illustrating the usage of Lemma 2.5. (**a**) Only one node n accesses s and $\text{dist}(n, n) \leq L$. (**b**) Two nodes n_1 and n_2 both accesses s with $\text{dist}(n_1, n_2) \leq L$ and $\text{dist}(n_2, n_1) \geq L$

 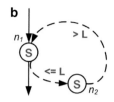

miss, but the position of s may move to smaller indices upon misses of n_1 (since $\text{dist}(n_2, n_1) > L$), which breaks down the memory block's movement monotonicity.

In order to use Lemma 2.5 to determine the k-**Miss** classification in the general case, we need to guarantee a global movement monotonicity of a memory block among all the related nodes. This can be done by examining the condition of Lemma 2.5 for all node pairs in a *strongly connected component* (maximal strongly connected subgraph) together, as described in the following theorem:

Theorem 2.1. *Let SCC be a strongly connected component in the CFG, let S be the set of nodes in SCC accessing the same memory block s. The total number of misses incurred by all the nodes in S is at most k if the following condition holds:*

$$\forall n_x, n_y \in S : \text{dist}(n_x, n_y) \leq k \tag{2.2}$$

where k is bounded by the number of cache lines L.

Proof. Let ev_f and ev_l be the first and last events triggered during program execution. Since S is a subset of the strongly connected component SCC, any event accessing s in the event sequence $\{ev_f, \cdots, ev_l\}$ has to be also triggered by some node in S (otherwise there will be a cycle including nodes both inside and outside SCC, which contradicts with that SCC is a strongly connected component).

By $k \leq L$, Condition (2.2) and Lemma 2.5, we know that among the events $\{ev_f, \cdots, ev_l\}$ whenever the access to s is a miss, s will be installed to a cache line with a strictly larger index than before. Since every time after s is accessed in the cache (either hit or miss), the corresponding MRU-bit is 1, so by Condition (2.2) and Lemma 2.3 we further know that among the events $\{ev_f, \cdots, ev_l\}$, as soon as s is installed to a cache line with index equal to or larger than k, it will not be evicted. In summary, there are at most k misses of s among events $\{ev_f, \cdots, ev_l\}$, i.e., the nodes in S have at most k misses in total.

2.5.3 Efficient k-**Miss** Determination

Theorem 2.1 gives us the condition to identify k-**Miss** node sets. The major task of checking this condition is to calculate the maximal stack distance $\text{dist}()$. As mentioned in Sect. 2.4, the exact calculation of $\text{dist}()$ is very expensive, which is the reason why the analysis of **LRU** relies on AI to obtain an over-approximate classification. For the same reason, we also resort to over-approximation to efficiently check the conditions of k-Miss. *The main idea is to use the analysis result for the same program under LRU to infer the desired k-Miss classification under MRU.*

Lemma 2.6. *Let n_y be a node that accesses memory block s and is classified as AH/FM by Must/Persistence analysis with a k-way LRU cache. For any node n_x that also accesses s, if there exists a cycle in the CFG including n_x and n_y, then the following must hold:*

$$\mathsf{dist}(n_x, n_y) \le k.$$

Proof. We prove the lemma by contradiction. Let n_x be a node that also accesses s and there exists a cycle in the CFG including n_x and n_y. We assume that $\mathsf{dist}(n_x, n_y) > k$. Then by the definition of $\mathsf{dist}(n_x, n_y)$ we know that there must exist a path p from n_x to n_y satisfying (i) $\pi(p) > k$ and, (ii) no other node accesses s apart from the first and last node along this path (otherwise $\mathsf{dist}(n_x, n_y) = 0$). This implies that under LRU, whenever n_y is reached via path p, s is not in the cache. Furthermore, n_y can be reached via path p repeatedly since there exists a cycle including n_x and n_y. This contradicts with that n_y is classified as AH/FM by the Must/Persistence analysis with a k-way LRU cache (Must/Persistence yields *safe* classification, so in the real execution an AH node will never be miss and an FM node can be miss for at most once).

Theorem 2.2. *Let SCC be a strongly connected component in the CFG, and S the set of nodes in SCC that access the same memory block s. If all the nodes in S are classified as AH by Must analysis or FM by Persistence analysis with a k-way LRU cache, then the node set S is k-Miss with an L-way MRU cache for $k \le L$.*

Proof. Let n_x, n_y be two arbitrary nodes in S, so both of them access memory block s and are classified as AH/FM by the Must/Persistence analysis with a k-way LRU cache. Since S is a subset of a strongly connected component, we also know n_x and n_y are included in a cycle in the CFG. Therefore, by Lemma 2.6 we know $\mathsf{dist}(n_x, n_y) \le k$. Since n_x, n_y are arbitrarily chosen, the above conclusion holds for any pair of nodes in S. Therefore, S can be classified as k-Miss according to Theorem 2.1.

Theorem 2.2 tells that we can identify k-Miss node sets with a particular k by doing Must/Persistence analysis with a LRU cache of the corresponding number of ways. Actually, we only need to do the Must and Persistence analysis once with an L-way LRU cache, to identify k-Miss node sets with *all* different k ($\le L$). This is because the Must and Persistence analysis for LRU cache maintains the information about the maximal age of a memory block at certain point in the CFG, which can be directly transferred to the analysis result with any cache size smaller than L. For example, suppose by the Must analysis with an L-way LRU cache, a memory block s has maximal age of k before the access of a node n, then by the Must analysis with a k-way LRU cache this node n will be classified as AH. We will not recite the details of Must and Persistence analysis for LRU cache or explain how the age information is maintained in these analysis procedures, but refer interested readers to the references [19, 110].

Moreover, the maximal age information in the Must and Persistence analysis with an 2-way LRU cache can also be used to infer traditional AH and FM classification under MRU according to the relative competitiveness property between MRU and LRU [95]: an L-way MRU cache is 1-competitive relative to a 2-way LRU cache, so a Must (Persistence) analysis with a 2-way LRU cache can be used as a sound Must (Persistence) analysis with an L-way MRU cache. Therefore, if the maximal age of a node in a Must (Persistence) analysis with an L-way LRU cache is bounded by 2 ($L \geq 2$), then this node can be classified as AH (FM) with an L-way MRU cache. Adding this competitiveness analysis optimization helps us to easily identify AH nodes when several nodes in a row access the same memory block. For example, if a memory block (i.e., a cache line) contains two instructions, then in most cases the second instruction is accessed right after the first one, so we can conclude that the second node is AH with a 2-way LRU cache, and thus is also AH with an L-way MRU. Besides dealing with the above easy case, the competitiveness analysis optimization sometimes can do more for *set-associative* caches with a relatively large number of cache sets. For example, consider a path accessing 16 pairwise different memory blocks, and a set-associative cache of 8 sets. On average only 2 memory blocks on this path are mapped to each set, so competitiveness analysis may have a good chance to successfully identify some AH and FM nodes.

2.5.4 Generalizing k-Miss for Nested Loops

Precisely predicting the cache behavior of loops is very important for obtaining tight WCET estimations. In this chapter, we simply define a loop $\mathcal{L}_\ell$ as a *strongly connected subgraph* in the CFG.[3] (Note the difference between a strongly connected *subgraph* and a strongly connected *component*.)

The ordinary CHMC may lead to over-pessimistic analysis when loops are nested. For example, Fig. 2.9 shows a program containing two-level nested loops and its (simplified) CFG. Suppose the program executes with a 4-way LRU cache. Since $\text{dist}(n_s, n_s) = 6 > 4$ (see $s \rightarrow f \rightarrow d \rightarrow e \rightarrow g \rightarrow b \rightarrow d \rightarrow s$), the memory block s can be evicted out of the cache repeatedly, and thus we have to put n_s into the negative classification according to the ordinary CHMC, and treat it as being always miss whenever it is accessed. However, by the program semantics we know that every time the program enters the inner loop it will iterate for 100 times,

---

[3]In realistic programs, loop structures are usually subject to certain restrictions (e.g., a *natural* loop has exactly one header node which is executed every time the loop iterates, and there is a path back to the header node [120]). However, the properties presented in this section are not specific to any particular structure, so we define a loop in a more generic way.

```
void main() {

    int i, j, x;

    for (i = 0; i++; i < 5 )

        for (j = 0; j++; j < 100 )

            x++;
}
```

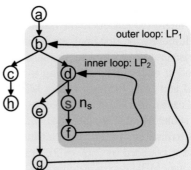

Fig. 2.9 A program with nested loop and its (simplified) CFG

during which s will not be evicted out of the cache since the inner loop can be fit into
the cache entirely. So node n_s has only 5 misses out of the total 500 cache accesses
during the whole program execution. Putting n_s into the negative classification and
treating it as being always miss is obviously over-pessimistic.

To solve this problem, [24, 106] reloaded the **FM** classification by relating it to
certain loop scopes:

Definition 2.6 (FM Regarding a Loop). A node is **FM** regarding a loop $\mathfrak{L}_\ell$ iff it
has at most one miss (at the first access) and otherwise will be always hit when the
program executes inside $\mathfrak{L}_\ell$.

In the above example node n_s is **FM** regarding the inner loop $\mathfrak{L}_2$.

The same problem also arises for **MRU**. Suppose the program in Fig. 2.9 runs
with a 4-way **MRU** cache. For the same reason as under **LRU**, node n_s has to be put
into the negative classification category. However, we have $\mathrm{dist}(n_s, n_s) = 3$ if only
looking at the inner loop, which indicates that n_s can be miss for at most 3 times
every time it executes inside the inner loop. As with **FM**, we can reload the k-**Miss**
classification to capture this locality:

Definition 2.7 (k-Miss Regarding a Loop). A node is k-**Miss** regarding a loop $\mathfrak{L}_\ell$
of the CFG iff it has at most k misses and all the other accesses are hits when the
program executes inside $\mathfrak{L}_\ell$.

The sought k-**Miss** classification under **MRU** for a loop can be inferred from
applying the **FM** classification under **LRU** to the same loop:

Theorem 2.3. *Let $\mathfrak{L}_\ell$ be a loop in the CFG, and S the set of nodes in the loop that
access the same memory block s. If all the nodes in S are classified as* **FM** *regarding*
$\mathfrak{L}_\ell$ *with a k-way* **LRU** *cache ($k \leq L$), then the node set S is k-**Miss** regarding $\mathfrak{L}_\ell$
with an L-way* **MRU** *cache.*

Proof. Similar to the proof of Theorems 2.1 and 2.2.

A node may be included in more than one k-Miss node sets regarding different loops. This typically happens across different levels in nested loops. For example, if the program in Fig. 2.9 executes with an 8-way MRU cache, then by Theorem 2.3 $\{n_s\}$ is classified as 3-Miss regarding the inner loop and 6-Miss regarding the outer loop. The miss number constraints implied by k-Miss with different k and different loops are generally incomparable. For example, with the loop bound setting in Fig. 2.9, 3-Miss regarding the inner loop allows at most $3 \times 5 = 15$ misses during the whole execution, which is "looser" than the outer loop 6-Miss which allows at most 6 misses. However, if we change the outer loop bound to 1, then the inner loop 3-Miss actually poses a "tighter" constraint as it only allows 3 misses while the outer loop 6-Miss still allows 6 misses. Although it is possible to explore program structure information to remove redundant k-Miss, we simply keep all the k-Miss classifications in our implementation since the ILP solver for path analysis can automatically and efficiently exclude such redundancy, as we illustrate in the next section.

2.5.5 WCET Computation by IPET

By now we have obtained the cache analysis results for MRU:

- k-Miss node sets that are identified by Theorems 2.2 and 2.3.
- AH and FM nodes that are identified using the relative competitiveness property between MRU and LRU as stated at the end of Sect. 2.5.3.
- All the nodes not included in the above two categories are NC.

Note that a node classified as AH by the relative competitiveness property may also be included in some k-Miss node set. In this case, we can safely exclude this node from the k-Miss node set, since AH provides a strong guarantee and the total number of misses incurred by other nodes in that k-Miss set is still bounded by k.

In the following we present how to apply these results in the path analysis by IPET to obtain the WCET estimation. The path analysis adopts a similar ILP formulation framework to the standard, but it is extended to handle k-Miss node sets. All the variables in the following ILP formulation are non-negative, which will not be explicitly specified for simplicity of presentation.

To obtain the WCET, the following maximization problem is solved:

$$Maximize \ \left\{ \sum_{\forall b_a} c_a \right\}$$

where c_a denotes the overall execution cost of basic block b_a (on the worst-case execution path). Since a basic block typically contains multiple nodes with different CHMC, the execution cost for each basic block is further refined as follows.

We assume the execution delay inside the processing unit is constant for all nodes, and the total execution delay of a node only differs depending on whether the cache access is a hit or a miss: C^h upon a hit and C^m upon a miss. Since the accesses of an AH node are always hits, the overall execution delay of an AH node n_i in b_a is simply $C^h \times x_a$ where the variable x_a represents the execution count of b_a. Similarly, the overall execution delay of an NC node is $C^m \times x_a$. The remaining nodes are the ones included in some k-Miss node sets (regarding some loops). For each of such nodes n_i, we use variables z_i ($\leq x_a$) to denote the execution count of n_i with cache access being miss. So the overall execution delay of a node n_i in some k-Miss node set is $C^m \times z_i + C^h \times (x_a - z_i)$. Putting the above discussions together, we have the total execution cost of a basic block b_a:

$$c_a = (\pi_{\mathsf{AH}} \times C^h + \pi_{\mathsf{NC}} \times C^m) \times x_a + \sum_{n_i \in b_a^*} \left(C^m \times z_i + C^h \times (x_a - z_i) \right)$$

where π_{AH} and π_{NC} is the number of AH and NC nodes in b_a, respectively, and b_a^* is the set of nodes in b_a that are contained in some k-Miss node sets (regarding some loops). Since at most k misses are incurred by a k-Miss node set regarding a loop $\mathcal{L}_\ell$ every time the program enters and iterates inside the loop, we have the following constraints to bound z_i:

$$\forall (S, \mathcal{L}_\ell) \text{ s.t. } S \text{ is } k\text{-Miss regarding } \mathcal{L}_\ell : \sum_{n_i \in S} z_i \leq k \times \sum_{e_j \in \mathsf{entr}_\ell} y_j$$

where entr_ℓ is the set of edges through which the program can enter $\mathcal{L}_\ell$ and we use variable y_j to denote how many times an edge $e_j \in \mathsf{entr}_\ell$ is taken during program execution. Recall that a node may be contained by multiple k-Miss sets (e.g., k-Miss regarding both the inner and outer loop with different k), so each z_i may be involved in several of the above constraints.

Besides the above constraints, the formulation also contains *program structural constraints* which are standard components of the IPET encoding. The WCET of the program is obtained by solving the above maximization problem, and the execution count for each basic block along the worst-case path is also returned.

2.6   Experimental Evaluation

The main purpose of the experiments is to evaluate

1. the precision of our proposed MRU analysis, and
2. the predictability comparison between LRU and MRU.

To evaluate (1), we compare the estimated WCET obtained by our MRU analysis and the measured WCET obtained by simulation with MRU caches. To evaluate (2), we compare the estimated WCET obtained by our MRU analysis and that by the state-of-the-art LRU analysis based on abstract interpretation (Must and May analysis in [19] and Persistence analysis in [110]). The smaller is the difference between the estimated WCET by our MRU analysis and by the LRU analysis, the more confident we are to claim that MRU is also a good candidate for cache replacement policies in real-time embedded systems, especially taking into account MRU's other advantages in hardware cost, power consumption, and thermal output.

2.6.1 Experiment Setup

As presented in Sect. 2.5.5, we assume the execution delay of each node only differs depending on whether the cache access is a hit or miss. The programs execute with a $1K$ bytes set-associative instruction cache. Each instruction is 8 bytes, and each cache line (memory block) is 16 bytes (i.e., each memory block contains two instructions). All instructions have a fixed latency of 1 cycle. The memory access penalty is 1 cycle upon a cache hit, and 10 cycles upon a cache miss. To conduct experiments with cache of different number of ways, we keep the total cache size fixed and change the number of cache sets correspondingly. Although the experiments in this chapter are conducted with *instruction* caches, the theoretical results of this work also directly apply to data caches, and we leave the evaluation for data caches as our future work.

The programs used in the experiments are from the Mälardalen Real-Time Benchmark suite [121]. Some programs in the benchmark are not included in our experiments since the CFG construction engine (from Chronos [122]) used in our prototype does not support programs with particular structures like recursion and switch-case very well. The loop bounds in the programs that cannot be automatically inferred by the CFG construction engine are manually set to be 50. The size of these programs used in our experiments ranges from several tens to about 4000 lines of C code, or from several tens to about 8000 assembly instructions compiled by a gcc compiler re-targeted to the SimpleScaler simulator [123] with $-O0$ option (no optimization is allowed in the compilation).

Since the benchmark programs have been compiled by a gcc compiler re-targeted to SimpleScalar, a straightforward way of doing the simulation is to execute the compiled binary on SimpleScalar (configured and modified to match our hardware configuration). However, the comparison between the measured execution time by this approach and the estimated WCET may be meaningless to evaluate the quality of our MRU analysis since (a) simulations may only cover program paths that are much "shorter" than the actual worst-case path, and (b) the precision of the estimated WCET also depends on other factors, e.g., the tightness of the loop bounds, which is out of the interest of this chapter. In other words, the estimated WCET can be always significantly larger than the measured execution time obtained by the above approach, regardless the quality of the cache analysis.

In order to provide meaningful quality evaluation of our MRU cache analysis, we built an in-house simulator, which is driven by the worst-case path information extracted from the solution of the IPET ILP formulation and only simulates the cache update upon each instruction. This enables us to get closer to the worst-case path in the simulation and exclude effects of other factors orthogonal to the cache behavior. Note that the solution of the IPET ILP formulation only restricts how many times a basic block executes on the worst-case path, which allows the flexibility of arbitrarily choosing among branches as long as the execution counts of basic blocks still comply with the ILP solution. In order to obtain execution paths that are as close to the worst-case path as possible, our simulator always takes different branches alternatively which leads to more cache misses. The manual and source code of the simulator are online available [124].

2.6.2 Results and Discussions

Tables 2.1 and 2.2 show the simulation and analysis results with 4-way caches. In simulation with each cache, for each program we record the measured execution time (column "sim. WCET") and the number of hits and misses. In the analysis with each cache, for each program we record the estimated WCET (column "est. WCET") and the number of memory accesses that can and cannot be classified as hit (column "hit" and "miss") respectively. We calculate the over-estimation ratio of the LRU and MRU analysis respectively (column "over est."). For example, the "sim. WCET" and "est. WCET" of program bs under LRU is 3911 and 3947, respectively, then the over-estimation ratio is $(3947 - 3911)/3911 = 0.92\%$. Finally, we calculate the excess ratio of MRU analysis over LRU analysis (column "exc. LRU"). For example, the estimated WCET of program bs under LRU and MRU is 3947 and 4089, respectively, then the excess ratio is $(4089 - 3947)/3947 = 3.60\%$.

The results show that the WCET estimation with our MRU analysis has very good precision: the over-estimation comparing with the simulation WCET is on average 2.06 %. We can also see that the estimated WCETs with MRU and LRU caches are very close: the difference is 1.17 % on average.

For several benchmark programs, the simulated WCETs are exactly the same under LRU and MRU. The reason is that MRU is designed to imitate the LRU policy with a cheaper hardware logic. In some cases, the cache miss/hit behavior under MRU could be exactly the same as that under LRU, and thereby we may obtain exactly the same simulated WCET with MRU and LRU for some programs. Moreover, the total number of memory accesses in the simulation may be different with two policies for the same program. This is because our simulator simulates the program execution with each policy according to the "worst-case" path information obtained from the solution of the corresponding ILP formula for WCET calculation. Sometimes, the ILP solutions with these two policies may correspond to different paths in the program, which may lead to different total numbers of memory accesses.

Table 2.1 Experiment results with 4-way caches

Program	Policy	Simulation			Analysis			Over Est. (%)	Exc. LRU (%)
		Hit	Miss	Sim. WCET	Hit	Miss	Est. WCET		
adpcm	LRU	1161988	56622	2946818	1158440	60170	2978750	1.08	
	MRU	1162890	55920	2490900	1155008	63802	3011838	2.41	1.11
bs	LRU	1741	39	3911	1737	43	3947	0.92	
	MRU	1740	39	3909	1720	59	4089	4.61	3.60
bsort	LRU	146781	69	294321	146773	77	294393	0.02	
	MRU	146781	69	294321	146718	132	294888	0.19	0.17
cnt	LRU	198496	103	398125	198489	110	398188	0.02	
	MRU	198496	103	398125	198443	156	398602	0.12	0.10
crc	LRU	104960	222	212362	104947	235	212479	0.06	
	MRU	104947	227	212391	104759	415	214083	0.80	0.75
edn	LRU	8506404	395838	21367026	8503690	398552	21391452	0.11	
	MRU	8506398	395844	21367080	8413450	488792	22203612	3.92	3.80
expint	LRU	43633	65	87981	43627	71	88035	0.06	
	MRU	43633	65	87981	43530	168	88908	1.05	0.99
fdct	LRU	15269	15421	200169	15268	15422	200178	<0.01	
	MRU	15269	15421	200169	15268	15422	200178	<0.01	0
fibcall	LRU	1009	27	2315	1006	30	2342	1.17	
	MRU	1009	27	2315	1006	30	2342	1.17	0
fir	LRU	58034	74	116882	58029	79	116927	0.04	
	MRU	58028	74	116870	57942	160	117644	0.66	0.61
insertsort	LRU	128844	53	258271	128841	56	258298	0.01	
	MRU	128844	53	258271	128811	86	258568	0.12	0.10

(continued)

Table 2.1 (continued)

Program	Policy	Simulation			Analysis			Over Est. (%)	Exc. LRU (%)
		Hit	Miss	Sim. WCET	Hit	Miss	Est. WCET		
janne	LRU	60794	37	121995	60788	43	122049	0.04	
	MRU	60793	37	121993	60779	51	122119	0.10	0.06
jfdctint	LRU	17302	15540	205544	17299	15543	205571	0.01	
	MRU	17302	15540	205544	17290	15552	205652	0.05	0.04
matmult	LRU	6331762	130	12664954	6331737	155	12665179	<0.01	
	MRU	6331760	132	12664972	6331606	286	12666358	0.01	0.01
minver	LRU	21841458	9655	43789121	21840200	10913	43800443	0.03	
	MRU	21841102	10011	43792325	21827892	23221	43911215	0.27	0.25
ndes	LRU	968253	20448	2161434	958951	29750	2245152	3.87	
	MRU	968333	20262	2159548	947654	40941	2345659	8.62	4.48
ns	LRU	245408124	77	490817095	245408119	82	490817140	<0.01	
	MRU	245408124	77	490817095	245408075	126	490817536	<0.01	<0.01
nsichneu	LRU	198708	198858	2584854	195706	201860	2611872	1.05	
	MRU	198708	198858	2584854	195706	201860	2611872	1.05	0
prime	LRU	3,617	63	7,927	3,585	95	8,215	3.63	
	MRU	3617	63	7927	3574	106	8.314	4.88	1.21
qsort	LRU	4209245	63	8419183	4206704	2604	8442052	0.27	
	MRU	4209245	63	8419183	4205204	4104	8455552	0.43	0.16
qurt	LRU	8417	250	19584	8341	326	20268	3.49	
	MRU	8432	235	19449	8227	440	21294	9.49	5.06

Table 2.2 Experiment results with 4-way caches—continued

Program	Policy	Simulation			Analysis			Over Est. (%)	Exc. LRU (%)
		Hit	Miss	Sim. WCET	Hit	Miss	Est. WCET		
select	LRU	3931319	7921	7949769	3930818	8422	7954278	0.06	
	MRU	3931319	7921	7949769	3927618	11622	7983078	0.42	0.36
sqrt	LRU	2990	52	6552	2984	58	6606	0.82	
	MRU	2986	52	6544	2936	102	6994	6.88	5.87
statemate	LRU	15976	17781	227543	15570	18187	231197	1.02	
	MRU	15926	17831	227993	15565	18192	231242	1.43	0.02
ud	LRU	11683773	5472	23427738	11683266	5979	23432301	0.02	
	MRU	11683765	5480	23427810	11671443	17802	23538708	0.47	0.45
Average	LRU							0.78	
	MRU							2.06	**1.17**

Then we conduct experiments with 8-way and 16-way caches (with the same total cache size but different number of cache sets). Note that it is rare to see set-associative caches with more than 16 ways in embedded systems, since a large number of ways significantly increase hardware cost and timing delay but brings little performance benefit [81]. So we did not conduct experiments with caches with more than 16 ways. Figure 2.10 summarizes the results with 8-way and 16-way caches, where the WCETs are normalized as the ratio versus the simulation results

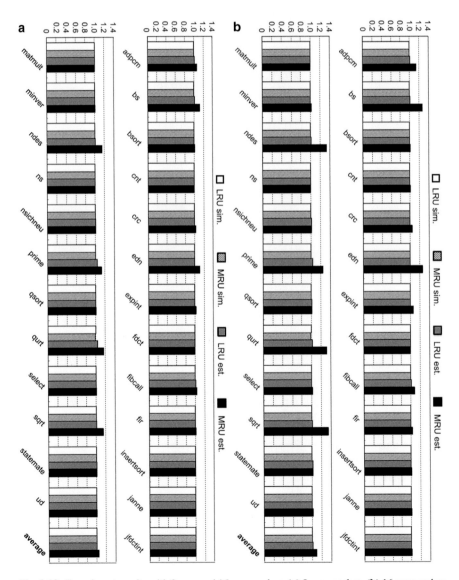

Fig. 2.10 Experiment results with 8-way and 16-way caches. (**a**) 8-way caches, (**b**) 16-way caches

under LRU. The over-estimation by our MRU analysis is **4.59** and **9.81** % for 8-way and 16-way caches, respectively, and the difference between the MRU and LRU analysis is 3.56 and 8.38 %. Overall, our MRU analysis still provides quite good precision on 8-way and 16-way caches.

We observe that for most programs the over-estimation ratio of the WCET by our MRU analysis scales about *linearly* with respect to the number of ways, the reason of which can be explained as follows. The k times of misses of k-Miss nodes is merely a theoretical bound for extreme worst-cases. In the simulation experiments, we observe that it hardly happens that a k-Miss node really encounters k times of misses. Most k-Miss nodes actually only incur one miss and exhibit similar behavior to FM nodes under LRU. For example, suppose a loop that contains k nodes accessing different memory blocks executes with k-way caches. Under LRU, the maximal ages of these nodes are all k, so our MRU analysis will be classified each of these nodes as k-Miss, and $k \times k = k^2$ misses have to be taken into account for the WCET estimation. However, in the simulation these k nodes can be entirely fit into the cache, and each of them typically only incurs one miss, so the number of misses reflected in the simulation WCET is typically k, which is k times smaller than that claimed by the analysis. So the ratio of over-estimated misses increases linearly with respect to the number of cache ways, and thus the over-estimation ratio in terms of WCET also scales about linearly with respect to the number of cache ways.

In the above experiments, while our MRU analysis has a precision close to that of LRU analysis for most programs, it obtains relatively worse performance for several programs (*bs*, *edn*, *ndes*, *prime*, *qurt*, and *sqrt*). While various program structures may lead to pessimism in our MRU analysis, there is a common reason behind that phenomenon, which can be explained as follows. The precision of our MRU analysis is sensitive to the ratio between the k value of k-Miss nodes and the number of times for which the loops containing these nodes iterate. For example, suppose a node is classified as 6-Miss with respect to a loop under MRU. If this loop iterates for 10 times, then the total execution cost of this node is estimated by $11 \times 6 + 2 \times 4 = 74$, where 11 is the execution cost upon a miss, 6 is the number of misses of this node, 2 is the execution cost upon a cache hit, and 4 is the number of hits of this node. On the other hand, this node is an FM with respect to the same loop under LRU, and the total execution cost is $11 \times 1 + 2 \times 9 = 29$. The estimated execution cost under MRU is about 2.5 times of that under LRU. However, if this loop iterates for 100 times, the total execution cost of this node under MRU is $11 \times 6 + 2 \times 94 = 254$, which is only 1.2 times of that under LRU ($11 \times 1 + 2 \times 99 = 209$). The high precision of our MRU analysis relies on the big amount of hits predicted by k-Miss. If a program contains many k-Miss nodes with comparatively large k values but iterates for a small number of times, the estimated WCET by our MRU analysis is less precise. This implies that, from the predictability perspective, MRU caches are more suitable for programs with relatively "small" loops that iterate for a great amount of times, e.g., with large loop bounds or nested-loops inside.

Figure 2.11 shows comparisons among the LRU analysis, the state-of-the-art MRU analysis (competitiveness analysis) and our k-Miss-based MRU analysis with various combinations of different optimization. Each column in the figure represents

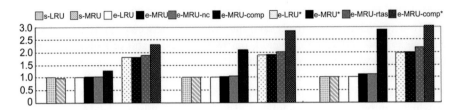

Fig. 2.11 Comparison of different analyses

the normalized WCET (the ratio versus the simulated WCET under LRU) averaged over all benchmark programs. With each cache setting, the first two columns are simulations, the next 4 columns are analyses *with* nested-loop optimization, and the last 4 columns are analyses *without* nested-loop optimization:

- **s-LRU**: Simulated WCET under LRU.
- **s-MRU**: Simulated WCET under MRU.
- **e-LRU**: Estimated WCET under LRU.
- **e-MRU**: Estimated WCET under MRU by the analysis in this paper.
- **e-MRU-nc**: Estimated WCET under MRU by the analysis in this chapter, but excludes the competitiveness analysis optimization.
- **e-MRU-comp**: Estimated WCET under MRU only by competitiveness analysis, which is the state-of-the-art MRU analysis before our k-Miss-based analysis.
- **e-LRU***: Estimated WCET under LRU but excludes the nested-loop optimization.
- **e-MRU***: Estimated WCET under MRU by the analysis in this paper but excludes the nested-loop optimization.
- **e-MRU-rtas**: Estimated WCET under MRU by the analysis in the previous conference version of this work [22].
- **e-MRU-comp***: Estimated WCET under MRU only by competitiveness analysis, but excludes the nested-loop optimization.

By comparing **e-MRU** with **e-MRU-comp** we can see that our new MRU analysis greatly improves the precision over the state-of-the-art technique for MRU analysis (competitiveness analysis), and the improvement is more significant as the number of cache ways increases. Recall that the competitiveness analysis relies on the analysis results for the same program with a 2-way LRU cache (with the number of cache sets unchanged, and thus the cache size scaled down to $\frac{2}{L}$ of the original L-way cache), so its results are more pessimistic when L is larger.

By the comparison among **e-MRU**, **e-MRU-nc**, **e-MRU***, and **e-MRU-rtas** we can see that both the competitiveness analysis and nested-loop optimization help to improve our MRU analysis precision. However, the contribution by the nested-loop optimization is much more significant.

By comparing columns 3–6 with columns 7–10 we see that in general adding nested-loop optimization can significantly improve the analysis precision. The only exception is **e-MRU-comp** with more cache ways (thus less cache sets, as we keep the total cache size unchanged), where even the memory blocks mapped to one cache set in an *inner* loop are too many to fit into 2 cache ways.

By comparing **e-MRU** with **e-LRU** and comparing **e-MRU-rtas** with **e-LRU***, we can see that the nested-loop optimization, which greatly affects the precision of each analysis, does not significantly affect the ratio between the estimated WCET under LRU and MRU. This is because our MRU analysis directly uses the LRU analysis results to find k-Miss nodes. With a more precise LRU analysis, our MRU analysis also becomes correspondingly more precise. This is why do this paper and its earlier conference version [22] draw similar conclusions about the precipitability comparison between LRU and MRU, although the analysis results in them are different.

We also evaluate the efficiency of our analysis. As presented in previous sections, our MRU analysis only requires to do the LRU cache analysis *once* to infer all the cache access classifications, so the MRU cache analysis procedure is as efficient as the state-of-the-art LRU cache analysis based on abstract interpretation. The interesting problem is the efficiency of the IPET-based path analysis, where more variables are used to support the constraints for k-Miss nodes. We solve the ILP formulation with an open source solver *lp_solve* [125] on a desktop machine with a 3.4 GHZ Core i7 2600 processor. The ILP formulation can be solved very efficiently: the calculation for each program takes on average 0.1 s and at most 0.8 s.

In summary, the experiment results show that our MRU analysis has both good precision and high efficiency. The estimated WCET by our MRU analysis is quite close to that by LRU analysis under common hardware setting, which indicates that MRU is a good candidate for cache replacement policies in real-time embedded systems, especially considering MRU's other advantages in hardware, power, and thermal efficiency.

2.7 Conclusions

This chapter studies the problem of WCET analysis with MRU caches. MRU was considered to be a very unpredictable replacement policy in the past, due to the lack of effective techniques to predict its hit/miss behavior. In this chapter, we disclose important properties of MRU, and develop efficient techniques to precisely bound the number of misses and thereby support high-quality WCET estimations with MRU caches. Experiments with benchmark programs indicate that the estimated WCET with MRU caches is rather close to that with LRU. This suggests a great potential for MRU to be used as the cache replacement policy in real-time embedded systems, especially considering the MRU's advantages in better cost, power, and thermal efficiency.

The experiments in this chapter only consider instruction caches. The reason is that our WCET analysis prototype does not support high-quality value analysis, so currently we cannot provide a meaningful evaluation with data caches. However, the properties of MRU disclosed in this chapter also hold for data caches, and our proposed analysis techniques can be directly applied to MRU data caches.

Chapter 3
FIFO Cache Analysis for WCET Estimation

Although most previous work in cache analysis for WCET estimation assumes the LRU replacement policy, in practice more processors use simpler non-LRU policies for lower cost, power consumption, and thermal output. This chapter focuses on the analysis of FIFO, one of the most widely used cache replacement policies. Previous analysis techniques for FIFO caches are based on the same framework as for LRU caches using *qualitative* always-hit/always-miss classifications. This approach, though works well for LRU caches, is not suitable to analyze FIFO and usually leads to poor WCET estimation quality. In this chapter, we propose a *quantitative* approach for FIFO cache analysis. Roughly speaking, the proposed quantitative analysis derives an upper bound on the "miss ratio" of an instruction (set), which can better capture the FIFO cache behavior and support more accurate WCET estimations. Experiments with benchmarks show that our proposed quantitative FIFO analysis can drastically improve the WCET estimation accuracy over previous techniques (the average over-estimation ratio is reduced from around 70 to 10 % under typical setting).

3.1 Introduction

A fundamental problem in the design and analysis of hard real-time systems is to bound the worst-case execution time (WCET) of programs [4]. To derive safe and tight WCET bounds, the analysis must take into account the cache architecture of the target processor. However, the cache analysis problem of statically determining whether each memory access is a hit or a miss is a challenging problem.

In the last two decades, precise and efficient analysis techniques have been developed for caches with a particular replacement policy, LRU (Least-Recently-Used). In contrast, less work has been done for other policies like MRU [97], FIFO [21], and PLRU [96]. However, in practice it is more common for commercial

© Springer International Publishing Switzerland 2016
N. Guan, *Techniques for Building Timing-Predictable Embedded Systems*,
DOI 10.1007/978-3-319-27198-9_3

processors to use non-LRU caches, which are simpler in hardware implementation but still have almost as good average-case performance as LRU [82]. Therefore, hardware manufacturers tend to choose these non-LRU policies, especially for embedded processors that are subject to strict cost, power, and thermal constraints.

This chapter studies the analysis of FIFO (First-In-First-Out), a cache replacement policy that is widely adopted in processor architectures like Intel XScale, ARM9, ARM11 [26]. The FIFO policy is very simple, but analyzing it is much harder than analyzing LRU. The state-of-the-art cache analysis techniques for WCET estimation are based on *qualitative* memory access classifications: to determine whether the memory accesses related to a particular instruction are *always* hits or *always* misses. Such an approach is highly effective for LRU caches since most instructions under LRU indeed exhibit such a "black or white" behavior. However, many instructions under FIFO exhibit a more nuanced behavior: a portion of the accesses are misses while all the other accesses are hits (e.g., at most $1/3$ of the accesses are misses). By existing analysis techniques based on the qualitative classification, such a behavior has to be treated as if these accesses are all misses, which inherently leads to very pessimistic analysis results. Recently, Grund and Reineke have developed FIFO analysis techniques based on the qualitative classification [21, 28]. Although their techniques are rather sophisticated, the derived WCET bounds are still grossly over-pessimistic (as shown in Sect. 3.6).

In this chapter we propose a *quantitative* approach to analyze FIFO caches, by which we can better capture the FIFO cache behavior and thus obtain much tighter WCET bounds for common programs. The proposed analysis derives an upper bound on the number of misses an instruction (set) may encounter through the whole program execution. As an efficient implementation, we use the cache analysis results of the same program under LRU replacement to derive the quantitative miss bound under FIFO replacement. Therefore, our technique inherits the advantages in efficiency and precision from the state-of-the-art LRU analysis techniques based on abstract interpretation [19].

The proposed analysis is based on a general metric *miss distance* of the underlying cache, and thus applies to any replacement policy as long as the miss distance of the underlying cache is known. The miss distance metric also enables an efficient *persistence* analysis to determine instructions that only encounter a *cold miss* but will always be hits afterwards, which further improves the overall analysis precision.

We have conducted experiments with benchmark programs on *instruction* caches to evaluate the quality of our proposed analysis. Experiments show that the estimated WCET by our FIFO analysis is much tighter than previous techniques (the average over-estimation ratio is reduced from around 70 to 10 % under typical setting), while still maintaining good analysis efficiency.

3.1.1 Relation to Previous Work

Although the always-hit/always-miss classification approach is dominating in previous work on cache analysis for WCET estimation [4, 126], recently there also have
been a couple of work towards the direction of quantitative cache analysis. Reineke
and Grund [101] studied the *relative competitiveness* between different policies by
providing upper (lower) bounds of the ratio on the number of misses (hits) between
two different replacement policies during the whole program execution. By this,
one can use cache analysis results under one replacement policy to predict the
number of cache misses (hits) of the same program under another policy. This
approach differs from our proposed quantitative cache analysis in several ways:
Firstly, while the relative competitiveness approach provides bounds on the number
of misses of the whole program, our quantitative cache analysis bounds the number
of misses at individual program points. Secondly, while the relative competitiveness
computation suffers scalability problems and thus does not cover cases with great
number of ways, our analysis can efficiently deal with large caches. Thirdly, the miss
(hit) bounds derived by the relative competitiveness is universal to all programs and
thus is much more pessimistic than our quantitative cache analysis in analyzing a
concrete program.

3.2 Preliminaries

3.2.1 Basic Concepts

For simplicity of presentation, we assume a *fully-associative* cache. However, the
analysis techniques of this chapter are directly applicable to *set-associative* caches,
since the accesses to memory references mapped to different cache sets do not
affect each other, and each cache set can be treated as a fully-associative cache and
analyzed independently. The memory content that fits into one cache line is called a
block.

Since this work focuses on the cache behavior, we do not consider the timing
effect of other components in the processor (e.g., pipeline and memory controller),
but assume the execution delay of each instruction only differs depending on
whether the cache access is a hit or a miss.

The program can be represented by a control-flow graph (CFG) $G = (N, E)$,
where $N = \{n_1, n_2, \cdots\}$ is the set of *nodes*, and $E = \{e_1, e_2, \cdots\}$ is the set of directed
edges. A *loop* $\mathfrak{L}$ in the CFG is a strongly connected subgraph of G. Note that here
we only provide a simple definition of the CFG and loops since the proposed cache
analysis does not rely on any particular CFG or loop structure. In Sect. 3.5 we will
redefine these definitions for the presentation of path analysis.

At run-time, when (a node of) the program accesses a block, the processor first checks whether the block is in the cache. If yes, it is a *hit*, and the program directly accesses this block from the cache. Otherwise, it is a *miss*, and this block is first installed in the cache before the program accesses it.

A block occupies only one cache line regardless how many times it is accessed. So the number of *different* blocks in an access sequence is important to the cache behavior. We use the following concept to reflect this:

Definition 3.1 (Stack Length). The stack length of an access sequence corresponding to a path p in the CFG, denoted by $\pi(p)$, is the number of different blocks accessed along p.

For example, the stack length of access sequence "$a \to b \to c \to a \to b$" is 3, since only a, b, and c are accessed.

3.2.2   LRU *and* FIFO

The cache update rule of LRU and FIFO is the same upon *misses*: when the program accesses a block δ that is not in the cache, all the blocks in the cache will be shifted one position to the next cache line (the block in the last cache line is removed from the cache), and δ is installed to the first cache line.

LRU and FIFO only differ in their update rules upon *hits*. Let the program access a block δ that is already in the cache. In LRU caches, δ is moved to the first cache line and all blocks that were stored before δ's old position will be shifted one position to the next cache line. In FIFO caches, δ stays at the original position and thus the whole cache keeps unchanged. Figure 3.1 illustrates the cache update upon an access to block δ on a 4-way LRU and FIFO cache, respectively.

3.2.3   LRU *Cache Analysis*

As mentioned in Sect. 3.1, our quantitative FIFO analysis uses the analysis results of the same program under LRU to infer the cache behavior under FIFO. Thus, we provide a brief review of the state-of-the-art LRU cache analysis technique.

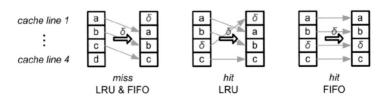

Fig. 3.1 Illustration of LRU and FIFO replacement

WCET estimation with precise cache analysis suffers from serious state-space explosion, so people resort to approximation techniques separating path analysis based on IPET (Implicit Path Enumeration Techniques) and cache analysis based on AI (Abstract Interpretation) for good scalability [19]. The AI-based LRU cache analysis uses three fix-point analyses on the abstract cache domain:

- Must analysis determines if the accesses of a node are always hits (AH);
- May analysis determines if the accesses of a node are always misses (AM);
- Persistence analysis determines if a node will at most encounter a cold miss and afterwards will be always-hit when the program executes inside a particular loop; the classification of such nodes is first-miss (FM) regarding the corresponding loop.

If a node is not determined by any of the above analyses, then it is classified as not-classified (NC). Under the problem model assumption of this chapter, NC nodes are treated in the same way as AM in the path analysis to calculate safe WCET bounds. We refer to the references [19, 106, 107, 110] for details about these fix-point analyses.

3.3   A New Metric: Miss Distance

This section introduces a general metric *miss distance*, which will be useful to establish the quantitative FIFO cache analysis in the next section. Before formally introducing the miss distance, we first use the following example to motivate why it is an interesting metric relevant to the timing predictability of cache replacement policies:

Given a loop accessing K blocks and a K-way cache. Since the whole loop can be fit into the cache, there is a strong intuition to claim the property that each node in the loop is FM regarding this loop. However, this is not always true. It depends on the underlying replacement policy: it holds for many policies including LRU, MRU, and FIFO, but not for others including PLRU.

This property is attractive since it enables a very efficient Persistence analysis by only counting the number of different blocks accessed in a loop. Since a program typically spends most of its execution time in loops, this property is highly relevant to the timing analysis of the whole program. Therefore, it is interesting to ask the following questions: What is the essence for a cache replacement policy to have this property? If it does not hold under a given policy, would it be true for a smaller loop? If yes, what is the upper limit of the loop size? Unfortunately, the existing cache replacement predictability metrics [26] cannot answer these questions.

Now we formally introduce the new metric *miss distance*:

Definition 3.2 (Miss Distance). The miss distance of a cache is the minimal number of different blocks being accessed between any pair of consecutive cache misses on the *same* block.

By examining the FIFO rule, it is easy to know:

Lemma 3.1. *The miss distance of a K-way FIFO cache is K.*

Proof. A block is installed to the first cache line upon a miss, and other K blocks need to be accessed to evict it.

The miss distance is K for a K-way LRU or MRU cache, and is $log_2K + 1$ for a K-way PLRU cache (proof omitted). LRU, MRU, and FIFO are optimal regarding this metric:

Lemma 3.2. *The miss distance of a K-way cache with any replacement policy is no larger than K.*

Proof. Assume a K-way cache with miss distance $K' > K$. Given a loop accessing K' different blocks, by Lemma 3.3 each of these blocks only encounters a cold miss and then be always hits when the program iterates inside the loop. However, this is impossible since a K-way cache cannot store more than K blocks at the same time.

With this new metric, we can answer the above questions:

Lemma 3.3. *Given a cache with miss distance X, and a loop $\mathfrak{L}$ in which the number of different blocks is no larger than X. Any block in $\mathfrak{L}$ encounters at most one miss (the cold miss) every time when the program executes inside $\mathfrak{L}$.*

Proof. Since the miss distance of the underlying cache is X, after the cold miss of a block δ, at least X other different blocks are needed for the next miss on δ to happen. However, this is impossible when the program executes inside the loop $\mathfrak{L}$ since it does not contain enough blocks.

Thus we have obtained a very efficient **Persistence** analysis: Given a K-way FIFO (LRU, MRU) cache, if the total number of different blocks accessed in loop $\mathfrak{L}$ is no larger than K, then all the nodes are **FM** regarding $\mathfrak{L}$. Similarly all the nodes in a loop accessing at most $log_2K + 1$ different blocks are **FM** regarding this loop on a K-way PLRU cache.

3.4   Quantitative FIFO Analysis

The idea behind the quantitative FIFO cache analysis is fairly simple. Consider the following access sequence:

$$\underline{\delta} \to a \to b \to \underline{\delta} \to c \to d \to \underline{\delta} \to e \to f \to g \to \underline{\delta} \to h \to i \to \underline{\delta}$$

Suppose the underlying FIFO cache has four ways, then by Lemmas 3.3 and 3.1 we know that for any pair of consecutive misses to δ there are at least 4 different blocks accessed in between. In the above sequence, if the first access to δ is a miss, then the second one must be a hit since only 2 blocks are accessed in between.

One can see that at most 3 out of the total 5 accesses to δ are misses. For any memory access sequence, one can calculate an upper bound on the misses for each block. However, there are exponentially many paths in the CFG and it is infeasible to do the above analysis for each individual path. In the following, we will show how to do the quantitative analysis in the context of the CFG structure, and in the next section the analysis result will be integrated into the IPET framework to efficiently calculate a WCET bound of the whole program. First define the *maximal stack distance* between two nodes accessing the same block in the scope of a certain loop:

Definition 3.3 (Maximal Stack Distance). Let n_i and n_j be nodes in loop $\mathfrak{L}$ accessing the same block δ (n_i and n_j may be the same node). The maximal stack distance from n_i to n_j regarding loop $\mathfrak{L}$, denoted by $\mathsf{dist}_{\mathfrak{L}}(n_i, n_j)$, is defined as:

$$\mathsf{dist}_{\mathfrak{L}}(n_i, n_j) = \begin{cases} \max\{\pi(p)|p \in P_{\mathfrak{L}}(n_i, n_j)\} & \text{if } P_{\mathfrak{L}}(n_i, n_j) \neq \emptyset \\ 0 & \text{otherwise} \end{cases}$$

where $P_{\mathfrak{L}}(n_i, n_j)$ is the set of paths satisfying:

- All nodes along the path are included in loop $\mathfrak{L}$;
- n_i (n_j) is the first (last) node of the path;
- No other nodes in the path, besides n_i and n_j, access δ.

Figure 3.2 illustrates the maximal stack distance related to block δ with an inner loop $\mathfrak{L}_{in}$ and an outer loop $\mathfrak{L}_{out}$, respectively. For example, we have $\mathsf{dist}_{\mathfrak{L}_{in}}(n_6, n_6) = 3$ since a "longest" path from n_6 back to n_6 in the scope of $\mathfrak{L}_{in}$ accesses 3 different blocks ($n_6 \rightarrow n_3 \rightarrow n_5 \rightarrow n_6$), while $\mathsf{dist}_{\mathfrak{L}_{out}}(n_6, n_6) = 6$ since the "longest" path in the scope of $\mathfrak{L}_{out}$ accesses 6 different blocks ($n_6 \rightarrow n_9 \rightarrow n_2 \rightarrow n_7 \rightarrow n_9 \rightarrow n_2 \rightarrow n_3 \rightarrow n_5 \rightarrow n_6$).

Lemma 3.4. *Given a cache with miss distance K. Let Δ be the set of nodes in a loop $\mathfrak{L}$ accessing block δ, and it holds*

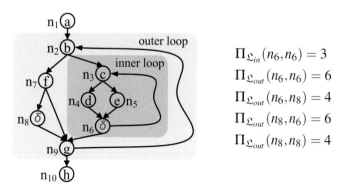

$$\Pi_{\mathfrak{L}_{in}}(n_6, n_6) = 3$$
$$\Pi_{\mathfrak{L}_{out}}(n_6, n_6) = 6$$
$$\Pi_{\mathfrak{L}_{out}}(n_6, n_8) = 4$$
$$\Pi_{\mathfrak{L}_{out}}(n_8, n_6) = 6$$
$$\Pi_{\mathfrak{L}_{out}}(n_8, n_8) = 4$$

Fig. 3.2 A CFG example. The letter inside each circle denotes the block accessed by this node

$$\forall n_i, n_j \in \Delta : \mathsf{dist}_{\mathfrak{L}}(n_i, n_j) \leq \ell \tag{3.1}$$

where ℓ is a positive integer no larger than K. Then the total number of misses for nodes in Δ is bounded by:

$$\lfloor \gamma \cdot x \rfloor + y \tag{3.2}$$

where $\gamma = 1/(1 + \lfloor (K-1)/(\ell-1) \rfloor)$, x is the total number of executions of nodes in Δ and y is the total number of times the program enters loop $\mathfrak{L}$ during the whole program execution.

Proof. The first step is to prove there are at least $\lfloor (K-1)/(\ell-1) \rfloor$ hits by nodes in Δ between any pair of consecutive misses for nodes in Δ. Since the miss distance of the underlying cache is K, after a miss of δ, at least K different blocks need to be accessed in order to evict δ from the cache. In other words, all the accesses to δ are hits, as long as the number of different blocks have been accessed after the first miss to δ does not exceed $K - 1$. By (3.1) we know that when the program executes inside loop $\mathfrak{L}$, the number of different blocks accessed between any two consecutive accesses to δ (not including δ) is at most $\ell - 1$. So δ will be accessed for at least $\lfloor (K-1)/(\ell-1) \rfloor$ times before it is evicted from the cache.

The program enters loop $\mathfrak{L}$ for y times. We use x_m ($1 \leq m \leq y$) to denote how many times δ is accessed when the program for the mth time enters and executes inside $\mathfrak{L}$. Above, we have proved there are at least $\lfloor (K-1)/(\ell-1) \rfloor$ hits by nodes in Δ between any pair of consecutive misses for nodes in Δ, so the total number of misses among these x_m accesses to δ can be bounded by:

$$1 + \lfloor x_m/(1 + \lfloor (K-1)/(\ell-1) \rfloor) \rfloor$$

Summing up this for each x_m we get an upper bound on the total number of misses for nodes in Δ:

$$y + \sum_{i=1}^{y} \lfloor x_m/(1 + \lfloor (K-1)/(\ell-1) \rfloor) \rfloor$$

By the general inequality property $\lfloor \frac{a}{c} \rfloor + \lfloor \frac{b}{c} \rfloor \leq \lfloor \frac{a+b}{c} \rfloor$ and $\sum_{i=1}^{y} x_m = x$, the above expression is bounded by (3.2).

Intuitively speaking, Lemma 3.4 implies a "ratio" γ of the misses over all the accesses by a set of nodes when the program iterates inside a loop. We call such a node set a *γ-set regarding* $\mathfrak{L}$. Note that a node may be included by several γ-sets regarding different loops and different γ values. For example, suppose the CFG in Fig. 3.2 is executed with a cache of miss distance 8, then n_6 is included in a singleton $\frac{1}{4}$-set regarding $\mathfrak{L}_{in}$ ($\gamma = 1/(1 + \lfloor (8-1)/(3-1) \rfloor) = 1/4$), as well as a $\frac{1}{2}$-set $\{n_6, n_8\}$ regarding $\mathfrak{L}_{out}$ ($\gamma = 1/(1 + \lfloor (8-1)/(6-1) \rfloor) = 1/2$).

To use Lemma 3.4, one needs to compute the maximal stack distance $\mathsf{dist}_{\mathfrak{L}}()$. In general, the time complexity of computing $\mathsf{dist}_{\mathfrak{L}}()$ is at least exponential regarding the number of cache ways,[1] so we need efficient approximation to handle real-life-size problems. Actually, computing $\mathsf{dist}_{\mathfrak{L}}()$ is exactly the essential problem to solve in the analysis of LRU caches. Therefore, we can use the over-approximate AI-based LRU analysis introduced in Sect. 3.2.3 to efficiently bound $\mathsf{dist}_{\mathfrak{L}}()$.

Lemma 3.5. *Given a ℓ-way* LRU *cache. Let Δ be the set of nodes in a loop $\mathfrak{L}$ accessing block δ. If all the nodes in Δ are classified as* AH *or* FM *regarding $\mathfrak{L}$ by the cache analysis, then it must hold:*

$$\forall n_i, n_j \in \Delta : \mathsf{dist}_{\mathfrak{L}}(n_i, n_j) \leq \ell \qquad (3.3)$$

Proof. Prove by contradiction. Assume two nodes n_i and n_j in S have $\mathsf{dist}_{\mathfrak{L}}(n_i, n_j) > \ell$. Then by the definition of $\mathsf{dist}_{\mathfrak{L}}(n_i, n_j)$ there is at least one path from n_i to n_j inside $\mathfrak{L}$ has stack length larger than ℓ. Now suppose this particular path is always taken when the program iterates inside the loop, then n_j will always encounter misses. This contradicts that the cache analysis claims that n_j is miss for at most once when the program executes inside this loop.

Now combining Lemmas 3.1, 3.4 and 3.5, we obtain the main result of this section:

Theorem 3.1. *Let Δ be the set of nodes in a loop $\mathfrak{L}$ accessing block δ. If all the nodes in Δ are classified as* AH *or* FM *regarding $\mathfrak{L}$ by a safe analysis on an ℓ-way* LRU *cache, then the total number of misses for nodes in Δ on a K-way* FIFO *cache is bounded by:*

$$\lfloor \gamma \cdot x \rfloor + y$$

where $\gamma = 1/\left(1 + \lfloor (K-1)/(\ell-1) \rfloor\right)$, x is the total number of executions of nodes in Δ and y is the total number of times the program enters loop $\mathfrak{L}$ during the whole program execution.

Since γ is non-decreasing with respect to ℓ, we want to find the minimal ℓ such that all nodes in Δ are classified as AH or FM regarding $\mathfrak{L}$ under LRU in order to minimize the "miss ratio." To do this, we actually only need to conduct the LRU cache analysis *once* with a K-way cache. This is because the Must and Persistence analysis for LRU maintains the information about the maximal age of a block at certain point in the CFG (when the program executes in a certain loop), which can be directly transferred to the analysis result with any LRU cache of size smaller than K. For example, suppose in the Must analysis with an 8-way LRU cache, a block

---

[1]This can be shown by a reduction from the well-known 3-SAT problem, the details of which are omitted due to the space limit.

δ has maximal age of 4 before the execution of a node accessing δ, then by the **Must** analysis with a 4-way **LRU** cache this node will be classified as **AH**. We will not recite the **LRU Must** and **Persistence** analysis details, neither explain how the age information is maintained in the analysis procedure. Interested readers can find details in the references [106, 107, 110].

3.5   Computation of WCET Bounds

In this section we introduce how to integrate the quantitative **FIFO** cache analysis results from the last section into IPET to efficiently compute a WCET bound of the analyzed program. First redefine the CFG on the basis of *basic blocks*:

Definition 3.4 (CFG). A CFG is a tuple $G = (B, E, b_{st})$:

- $B = \{b_1, b_2, \cdots\}$ is the set of *basic blocks* in the CFG;
- $E = \{e_1, e_2, \cdots\}$ is the set of directed *edges* connecting the basic blocks in the CFG;
- $b_{st} \in B$ is the unique *starting basic block* of the CFG.

As a common restriction in structured programming [127], we assume each loop contains a single *head basic block*, and the program can jump into the loop by reaching the head basic block via some *entry edges*. The *loop bound* restricts the maximal times the loop iterates every time the program enters it. The head basic block tests whether the loop condition is satisfied. If yes, the program continues to execute the *body basic blocks*, which are the basic blocks in the loop excluding the head basic block, otherwise the program exists the loop. Formally, a loop is defined as:

Definition 3.5 (Loop). A loop in the CFG is represented by a tuple $\mathcal{L}_l = (\mathsf{entr}_l, \mathsf{head}_l, \mathsf{body}_l, \mathsf{lpb}_l)$ with:

- entr_l: the set of entry edges of the loop;
- head_l: the head basic block of the loop;
- body_l: the set of all body basic blocks of the loop;
- lpb_l: the loop bound.

The overall **FIFO** cache analysis results can be summarized as follows: **AH** nodes decided by the **Must** analysis in [21, 28], **FM** nodes (regarding some loop) decided according to Lemma 3.3, and γ-sets (regarding some loop) determined by Theorem 3.1. Finally, the nodes that do not belong to any of the above classification are treated as **AM**. Note that if a node n_i is **FM** regarding loop $\mathcal{L}$, the number of misses with n_i is bounded by the number of times the program enters this loop, so $\{n_i\}$ can be viewed as a special case of γ-set with $\gamma = 0$ [the bound (3.2)] becomes $y + \lfloor 0 \cdot x \rfloor = y$). For simplicity of presentation, in the following we use term γ-set to include both the original ones derived by Theorem 3.1 and the **FM** singleton sets with $\gamma = 0$.

The standard IPET for WCET computation with LRU caches is encoded as an ILP (Integer Linear Programming) problem. Since our FIFO cache analysis results involve non-integers (miss ratio γ), we encode the IPET for FIFO cache as an MILP (Mixed-Integer Linear Programming) problem. The constants used in MILP formulation include C^h (the execution delay of each node upon a cache hit), C^m (the execution delay of each node upon a cache miss), and the miss ratio γ for each γ-set.

The formulation uses the following *non-negative* variables:

- c_a: for each b_a, c_a is b_a's total execution cost,
- x_a: for each b_a, x_a is the execution count of b_a,
- y_j: for each edge e_j, y_j counts how many times this edge is taken during the whole execution,
- z_i: for each node n_i included in some γ-set, z_i counts how many times n_i executes as cache misses.

The following maximization object is a safe WCET bound of the analyzed program:

$$Maximize \left\{ \sum_{\text{all } b_a} c_a \right\}$$

The following constraints are respected to bound the object.

Cost Constraints: The overall delay of an AH (AM) node n_i in b_a is simply $C^h \cdot x_a$ ($C^m \cdot x_a$). The remaining nodes are the ones included in some γ-set (including FM nodes as stated above). For each of such nodes n_i, we use variables z_i (s.t. $z_i \leq x_a$) to denote the execution count of n_i with cache accesses being misses. So the overall delay of such a node n_i is $C^m \cdot z_i + C^h \cdot (x_a - z_i)$. Putting the above discussions together, we have the total execution cost of each basic block:

$$\forall b_a : c_a = (\pi_{ah} C^h + \pi_{am} C^m) \cdot x_a + \sum_{n_i \in b_a^*} \left(z_i \cdot C^m + (x_a - z_i) \cdot C^h \right)$$

where π_{ah} and π_{am} is the number of AH and AM nodes in b_a, respectively, and b_a^* is the set of nodes in b_a that are involved in some γ-set.

γ-Set Constraints: The total number of misses for nodes in a γ-set regarding a loop $\mathfrak{L}_l$ is bounded by $\lfloor \gamma \cdot x \rfloor + y$, where x is the total number of executions of nodes in this γ-set and y is the total number of times the program enters $\mathfrak{L}_l$. So we can bound the number of misses incurred by a γ-set:

$$\forall (S, \mathfrak{L}_l) \text{ s.t. } S \text{ is a } \gamma\text{-set regarding } \mathfrak{L}_l :$$

$$\sum_{n_i \in S} z_i \leq \sum_{e_j \in \text{entr}_l} y_j + \sum_{n_i \in S} \lfloor x_a \cdot \gamma \rfloor$$

where entr_l is the set of entry edges of $\mathfrak{L}_l$ and y_j to denote how many times an edge $e_j \in \text{entr}_l$ is taken during the whole program execution. Recall that a node may be contained by multiple γ-sets, so each z_i may be involved in several of the above constraints.

Structure Constraints: The same as in the standard encoding of IPET with LRU caches [19].

3.6 Experimental Evaluation

We assume the execution delay of each node only differs depending on whether the cache access is a hit or a miss: all instructions have the same execution delay of 1 cycle, the memory access penalty is 1 cycle upon a cache hit and 10 cycles upon a cache miss. Each instruction is 8 bytes, and each block (cache line) is 16 bytes (i.e., each block contains two instructions). The programs used in the experiments are from the Mälardalen Real-Time Benchmark [121]. Some loop bounds cannot be automatically inferred, which are manually set to be 50. The size of these programs used in our experiments ranges from several tens to about 4000 lines of C code, or from several tens to about 8000 assembly instructions compiled by a gcc compiler re-targeted to the SimpleScalar [123] with -o0 option.

Simulation experiments are conducted with our in-house simulator, which is driven by the worst-case path information extracted from the solution of the MILP formulation. This approach can exclude the effects of other factors orthogonal to the cache behavior (e.g., the tightness of loop bounds), by which we can better evaluate the quality of the cache analysis itself than using traditional full-processor simulations. The solution of the MILP formulation only restricts how many times a basic block executes on the worst-case path, which allows the flexibility of arbitrarily choosing upon branches as long as the execution counts of basic blocks still comply with the MILP solution. In order to obtain execution paths that are as close to the worst-case path as possible, our simulator always takes different branches alternatively which leads to more cache misses.

Figure 3.3 shows the WCET estimations with a 0.5 K 4-way FIFO cache by the analysis of this work and Grund and Reineke's Must analyses [21, 28] (a node is

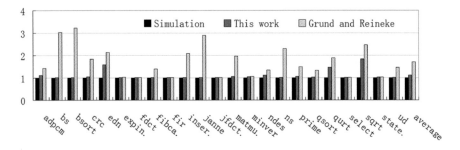

Fig. 3.3 Experiment results with a 0.5 K 4-way cache

classified as AH if it is determined by one of these two analyses). The must analyses in [21, 28] can only determine AH nodes. However, for a more fair comparison, we integrate these two must analyses with the VIVU technique [19] and thus they can also be used to determine FM nodes. The x-axis of the figure represents different benchmark programs (the last group is the average over all programs), and the y-axis is the *normalized* WCET estimation (the ratio between the WCET estimation and the execution time obtained by simulation). For most programs, the WCET estimation with our quantitative analysis is very close to simulation results: the normalized WCET estimation is on average 110.3 %. In contrast, the WCET estimation by Grund and Reineke's Must analysis is grossly pessimistic: the normalized WCET estimation is on average 171.8 %. In other words, by our new analysis, the over-estimation ratio is reduced from 71.8 to 10.3 %.

We also conducted experiments with various configurations: the cache size is 0.5 K, 1 K, or 2 K; the number of ways is 4, 8, 16, or 32 (the number of sets changes correspondingly, resulting in 12 different configurations). These experiments showed that our analysis is even more accurate with a greater cache size and/or a large number of cache ways, while the quality of Grund and Reineke's Must analysis is similar under different configurations. Detailed result figures with different configurations are omitted due to space limit.

Our FIFO analysis uses the analysis results of the same program under LRU to derive the quantitative guarantee, and thus is as efficient as the state-of-the-art LRU cache analysis based on abstract interpretation. The IPET with our quantitative FIFO analysis is encoded as an MILP problem and uses a greater number of variables, thus in general takes more time to solve than the standard ILP formulation in previous LRU analysis. However, some pragmatic optimizations in the MILP encoding are possible to improve the efficiency of the MILP solver performance.[2] Experiments showed that this approach has a good analysis efficiency with the benchmark programs in use. We solved the MILP formulation by lp_solve [128] on a laptop with an Intel Core i7 CPU (2.7 GHZ). The computation for each program takes at most several seconds.

3.7 Discussion and Conclusion

This chapter presented a quantitative approach for FIFO cache analysis. Unlike the previous standard cache analysis methods based on qualitative AH/AM classification, this new approach quantitatively bounds the number of misses caused by an instruction (set) during the whole program execution. Experiments with benchmark programs showed that the proposed analysis can significantly improve the WCET estimation accuracy over previous techniques while still maintains good efficiency.

---

[2]The idea is to group as many nodes with the same γ-set characterization into "blocks," to reduce the number of variables used in the MILP encoding.

Although the quantitative analysis approach supports a significantly better precision in predicting the *number* of cache hits/misses, we would like to point out that the guarantee provided by qualitative analysis is stronger and has the benefit of, e.g., being easier to be integrated in the analysis of architectures that are not fully timing-compositional [27]. Therefore, the next step of our work is to study how to integrate the quantitative cache analysis with the analysis of other components in the processor (e.g., pipelines). We also plan to evaluate the scalability of the proposed analysis with large-scale programs, and extend to multi-level caches and another widely used policy PLRU [23].

Appendix: The Complete IPET Formulation

We first introduce the loop structures adopted in our ILP encoding. As a common restriction in structured programming [127], we assume each loop contains a single *head basic block*, and the program can jump into the loop by reaching the head basic block via some *entry edges*. The *loop bound* restricts the maximal times the program iterates every time it enters the loop. The head basic block tests whether the loop condition is satisfied (e.g., the loop bound has not been reached). If the loop condition is satisfied, the program continues to execute the *body basic blocks*, which are the basic blocks in the loop excluding the head basic block, otherwise the program exists the loop. Formally we define a loop as:

Definition 3.6 (Loop Structure). A loop in the CFG is a tuple $\mathfrak{L}_\ell = ($entr$_\ell$, head$_\ell$, body$_\ell$, lpb$_\ell)$ with:

- entr$_\ell$: the set of entry edges of the loop;
- head$_\ell$: the head basic block of the loop;
- body$_\ell$: the set of all body basic blocks of the loop;
- lpb$_\ell$: the loop bound.

The ILP formulation uses the following constants

- C^h: the execution delay of each node upon a cache hit,
- C^m: the execution delay of each node upon a cache miss,

and the following *non-negative* variables

- c_a: for each basic block b_a, c_a is b_a's total execution cost,
- x_a: for each basic block b_a, x_a is the execution count of b_a,
- y_j: for each edge e_j in the entry edge set entr$_\ell$ of some loop $\mathfrak{L}_\ell$, y_j counts how many times this edge is taken during the whole execution,
- z_i: for each node n_i included in some k-Miss node sets regarding some loops, z_i counts how many times n_i executes with cache misses.

To obtain the WCET, the following maximization problem is solved:

$$\text{Maximize } \left\{ \sum_{\forall b_a} c_a \right\}$$

The following constraints are respected to bound the total cost:

- **Cost Constraint**: As discussed in Sect. 2.5.5, the total cost of each basic block is

$$\forall b_a : c_a = (\pi_{\mathsf{AH}} \times C^h + \pi_{\mathsf{NC}} \times C^m) \times x_a + \sum_{n_i \in b_a^*} \left(C^m \times z_i + C^h \times (x_a - z_i) \right)$$

where π_{AH} and π_{NC} is the number of AH and NC nodes in b_a, respectively, and b_a^* is the set of nodes in b_a that are contained in some k-**Miss** node sets (regarding some loops). Additionally, each $n_i \in b_a^*$ should satisfy $z_i \leq x_a$.

- **k-Miss Constraint**: As discussed in Sect. 2.5.5, the following constraints bound the number of misses incurred by a k-**Miss** node set:

$$\forall (S, \mathcal{L}_\ell) \text{ s.t. } S \text{ is } k\text{-}\mathsf{Miss} \text{ regarding } \mathcal{L}_\ell : \sum_{n_i \in S} z_i \leq k \times \sum_{e_j \in \mathsf{entr}_\ell} y_j$$

- **Structure Constraint**: Each basic block should have balanced input and output:

$$\forall b_a : x_a = \sum_{e_j \in \mathsf{input}(b_a)} y_j = \sum_{e_j \in \mathsf{output}(b_a)} y_j$$

The start basic block b_{st} of the program is executed only once:

$$x_{st} = 1$$

Each time the program enters the loop, each body basic block executes for at most lpb_ℓ times, so we have

$$\forall \mathcal{L}_\ell, \forall b_a \in \mathsf{body}_\ell : x_a \leq \mathsf{lpb}_\ell \times \sum_{e_j \in \mathsf{entr}_\ell} y_j$$

The head basic block may execute one more time to realize that the loop condition is not satisfied and thus the program exists the loop, so we have:

$$\forall \mathcal{L}_\ell, b_a = \mathsf{head}_\ell : x_a \leq (\mathsf{lpb}_\ell + 1) \times \sum_{e_j \in \mathsf{entr}_\ell} y_j$$

Part II
Real-Time Scheduling on Multicores

Chapter 4
Analyzing Preemptive Global Scheduling

Recently, there have been several promising techniques developed for schedulability analysis and response time analysis for multiprocessor systems based on over-approximation. In this chapter, we apply Baruah's window analysis framework [53] to response time analysis for sporadic tasks on multiprocessor systems to improve the analysis precision. The crucial observation is that for global fixed-priority scheduling, a response time bound of each task can be efficiently estimated by fixed-point computation without enumerating all the busy window sizes as in [53] for schedulability analysis. The technique is proven to dominate theoretically state-of-the-art techniques for response time analysis for multiprocessor systems. Our experiments also show that the technique results in significant performance improvement compared with several existing techniques for multiprocessor schedulability analysis.

4.1 Introduction

There is an increasing interest in developing real-time applications on multiprocessor platforms due to the broad introduction of multi-core chip processors. It is also one of the most challenging problems today for the real-time research community to develop efficient techniques for the analysis of such systems.

Recently, there have been several promising techniques developed for schedulability analysis, e.g., [53] and response time analysis, e.g. [129], for global multiprocessor scheduling. In this chapter, we take a second look at the problem of Response Time Analysis (RTA) for multiprocessor systems with global fixed-priority scheduling. We will present a new RTA technique to further improve the analysis precision of the existing techniques and also a non-trivial extension of the technique to task systems with arbitrary deadlines for multiprocessor systems.

© Springer International Publishing Switzerland 2016   71
N. Guan, *Techniques for Building Timing-Predictable Embedded Systems*,
DOI 10.1007/978-3-319-27198-9_4

Roughly speaking, RTA is to estimate the response time bound for each task in a set of tasks when they are scheduled using a given scheduling policy. It is an important technique for the design and analysis of not only hard real-time systems as it may be used for schedulability analysis but also soft real-time systems as the response time bounds provide an indication on how a system performs. For single processor systems, RTA has been intensively studied in the past two decades, and extended to various task models [35, 39, 41, 130, 131], to deal with real-life systems. Today we have obtained a rather good understanding of RTA for single-processor scheduling. In contrast, much less work on RTA for multiprocessor scheduling has been done by now.

Our work is mainly built upon the work of Bertogna and Cirinei [129] and Baruah [53]. We apply the window analysis framework of Baruah [53] to response time analysis inspired by the work of Bertogna and Cirinei [129] for sporadic tasks on multiprocessor systems with both constrained and arbitrary-deadline tasks. The crucial observation is that when the earliest time instant, after which all processors are occupied with tasks of higher priorities, occurs *just before* the release of a task, it results in an upper bound of the worst-case response time of the task for global fixed-priority scheduling. This allows us to efficiently compute the bound by fixed-point computation without enumerating all the busy window sizes as in [53] for schedulability analysis.

4.2 Problem Setting

We consider global fixed-priority scheduling on a multiprocessor platform consisting of M processors.

A sporadic task set τ consists of N sporadic tasks running on this platform. We use $\tau_i = \langle C_i, D_i, T_i \rangle$ to denote such a task where C_i is the *worst-case execution time* (WCET), D_i is the relative deadline for each release, and T_i is the minimum inter-arrival separation time also referred to as the *period* of the task. We further assume that all tasks are ordered by priorities, i.e., τ_i has higher priority than τ_j iff $i < j$. The *utilization* of a task τ_i is $U_i = C_i/T_i$.

A *constrained-deadline* task τ_i satisfies the restriction $D_i \leq T_i$, whereas an *arbitrary-deadline* task τ_i does not constrain the relation between D_i and T_i. We consider constrained-deadline tasks in this chapter.

A sporadic task τ_i generates a potentially infinite sequence of *jobs* with successive job-arrivals separated by at least T_i time units. We use J_i^h to denote the h-th job of τ_i. We omit the superscript h and just use J_i to denote a job of τ_i if there is no need to identify which job it is. Each job J_i^h adheres to the conditions C_i and D_i of its task τ_i and has additional properties concerning *absolute time points* related to its execution, which we denote with lowercase letters. The *release time* is denoted by r_i^h, the *deadline* by d_i^h which is derived by $d_i^h = r_i^h + D_i$, and the *finish time* by

f_i^h, which is the time instant at which J_i^h just finished its execution. We define the *response time* of J_i^h as the difference between its release and finish times:

$$R_i^h = f_i^h - r_i^h$$

The worst-case response time (WCRT) R_i of task τ_i is the maximal response time value among all jobs of τ_i in all job sequences possible in the system.

Since D_i is allowed to be larger than T_i, it is possible that several jobs of a task are *active* (i.e., released but not yet finished) simultaneously. We restrict that a job J_i^h can execute only if its precedent job J_i^{h-1} has been already finished, to avoid unnecessary working space conflict. This restriction is commonly adopted in the implementation of real-time operating systems for multi-cores/multiprocessors, for instance, RTEMS [33] and LITMUS^{RT} [34]. Thus, we define the *ready time* γ_i^h of J_i^h as $\gamma_i^h = \max(r_i^h, f_i^{h-1})$, which is the earliest time instant for a released job J_i^h to execute if no higher-priority task is interfering with it. At all time points $t \in [\gamma_i^h, f_i^h)$ we call J_i^h *ready*. Note that there is at most one ready job of each task at each time point (also for arbitrary deadlines).

We use the discrete time concept, i.e., any time value involved in the scheduling is a non-negative integer. This is based on the assumption that all events in the system happen only at clock ticks. Thus, we use time point t to represent the entire time interval $[t, t + 1)$.

Without loss of generality, we assume that tasks are strictly periodic (i.e., that $r_i^h = r_i^{h-1} + T_i$), unless stated otherwise. However, all the results are also applicable to sporadic task sets in general.

For simplicity of expression, we further use the following notations to express that a value A is "limited" if it is below or above a certain threshold B or C, respectively: $[A]_B = \max(A, B), [A]^C = \min(A, C)$, and $[A]_B^C = [[A]_B]^C$. This expression just keeps the value A if it is within the interval $[B, C]$, otherwise it is B if $A < B$ or C if $A > C$.

4.3 Previous Work

Before presenting our proposed techniques, we briefly review the previous work on RTA, to provide a primary knowledge background to readers that are not familiar with this field, as well as outline the contributions of this chapter against previous work.

4.3.1 The Basic Single-Processor Case

The RTA technique was for the first time proposed in [31], where it is only applicable to constrained-deadline task sets (i.e., $\forall \tau_i \in \tau : D_i \leq T_i$).

RTA of a task τ_k is based on the concept of *level-k busy period*. Intuitively, the *level-k busy period* is the maximum continuous time interval during which a processor executes tasks of priority greater than or equal to the priority of the considered task τ_k, until τ_k finishes its active job. For single-processor fixed-priority scheduling, the situation exhibiting the worst-case response time is known to happen at a well-defined *critical instant*: All higher priority tasks are released together with the analyzed task τ_k, i.e., at the same time instant. Thus, the maximal interference suffered by τ_k in a level-k busy period of length x can be computed by $\sum_{i<k} \lceil x/T_i \rceil \cdot C_i$. Using this, τ_k's WCRT can be calculated by finding the minimal solution of the following recursive equation:

$$ x = \sum_{i<k} \left\lceil \frac{x}{T_i} \right\rceil \cdot C_i + C_k $$

This solution can be found by interpreting the RHS as a monotonic function in x, whose minimal fixed point can be computed iteratively, starting at $x = C_k$.

4.3.2 The Basic Multiprocessor Case

RTA has been applied to multiprocessor scheduling with constrained-deadline task systems. The difference to the single-processor case is that the critical instant in multiprocessor scheduling is generally unknown. This prevents calculation of the exact interference suffered by the analyzed task τ_k during a level-k busy period. Instead, one has to derive an upper bound of the interference.

The work done by a task τ_i in the worst case during a level-k busy period can be divided into three parts:

- *body*: the contribution of all jobs (called *body jobs*) with both release time and deadline *in* the level-k busy period;
- *carry-in*: the contribution of at most one job (called *carry-in job*) with release time *earlier* than the level-k busy period and deadline *in* the level-k busy period;
- *carry-out*: the contribution of at most one job (called *carry-out job*) with release time *in* the level-k busy period and deadline *after* the level-k busy period.

A naive upper bound of the workload of each task τ_i during a level-k busy period of length x is obtained by assuming that the carry-in and carry-out of τ_i both contribute C_i each:

$$ W_k^{naive}(\tau_i, x) = \left\lceil \frac{x}{T_i} \right\rceil C_i + C_i $$

Adding the workload of all higher-priority tasks, one can use the term $\frac{1}{M} \sum_{i<k} W_k^{naive}(\tau_i, x)$ as an upper bound of the interference time suffered by τ_k during the level-k busy period of length x, due to all higher-priority tasks workload.

Therefore, an upper bound of the response time of τ_k can be obtained by finding the minimal solution of the following recursive equation [56, 132, 133]:

$$x = \frac{1}{M} \sum_{i<k} \left(\left\lceil \frac{x}{T_i} \right\rceil C_i + C_i \right) + C_k$$

Bertogna and Cirinei [129] have significantly improved the above result. We refer to their RTA technique as [BC-RTA] for short and give now a short overview of their key ideas. Firstly, rather than assuming that the carry-in and carry-out of a task are both C_i, they derived a more precise upper bound $W_k(\tau_i, x)$ of the workload for each task τ_i which is more precise than $W_k^{naive}(\tau_i, x)$, by carefully identifying the worst-case workload of each individual task.

Secondly, they observed that if the workload of a task τ_i is "too large," not necessarily all its workload can cause interference to the analyzed task τ_k, since the "extra" part of τ_i's workload has to be executed in parallel with τ_k. This is a fundamental difference between single-processor scheduling and multiprocessor scheduling (since in single-processor scheduling, no parallel execution takes place). In particular, they define the *interference* of τ_i to τ_k during a level-k busy period of length x as:

$$I_k(\tau_i, x) = [W_k(\tau_i, x)]_0^{x-C_k+1} \tag{4.1}$$

Using this observation, the recursive equation becomes:

$$x = \left\lfloor \frac{1}{M} \sum_{i<k} I_k(\tau_i, x) \right\rfloor + C_k \tag{4.2}$$

Note that in Eq. (4.1), the upper bound of τ_i's interference is $(x - C_k + 1)$ rather than $(x - C_k)$. With $(x - C_k)$ as the upper bound, the solution we get from Eq. (4.2) would not be guaranteed to be the upper bound of τ_k's response time. Intuitively, the "+1" is necessary to increase the right-hand side of (4.2) as long as there is more interference that could potentially prevent τ_k from running. For example, when the iterative search for the least fixed point is started with $x = C_k$, the search would stop immediately, since $\min(W_k(\tau_i, x), x - C_k)$ would be 0 for all i. A formal explanation of this issue can be found in [129].

4.4 Constrained-Deadline Task Sets

4.4.1 Busy Period Extension

In [BC-RTA], to derive a safe upper bound of the interference suffered by the analyzed task τ_k, it is assumed that every higher priority task τ_i has carry-in. This is an over-pessimistic assumption, since in a real scheduling sequence, it may be the case that some task τ_i's carry-in job has finished before the beginning of the busy period, therefore it actually does not contribute any carry-in to the interference of τ_k. To address a similar problem in schedulability tests of global EDF scheduling, Baruah [53] extends the busy period to an earlier time instant, which allows to bound the number of tasks doing carry-in by $M - 1$ (with M being the number of processors), which is in general much smaller than the N (the number of tasks) in [BC-RTA].

In the following we will apply Baruah's idea of busy period extension to RTA. It should be noted that a trivial combination of the busy period extension and RTA will lead to very high computational complexity, and could fail to yield analysis results for large-scale systems in reasonable time. However, as will be shown later, we can combine them without decreasing the analysis efficiency.

From now on, we let J_k be a job of τ_k that has the worst-case response time. As in [53], we extend the beginning of the level-k busy period from r_k (the release time of J_k) to an earlier time instant t_0, which is defined as the earliest time instant before r_k, such that at any time instant $t \in [t_0, r_k)$ all processors are occupied by tasks with higher priority than τ_k. If there is no such a time instant, we set $t_0 = r_k$. Using this, the *level-k busy period* is defined as the time interval $[t_0, f_k)$, and we define $\varphi = r_k - t_0$, which is the time span by which the busy period extends to the left, as shown in Fig. 4.1.

This definition of t_0 is chosen to impose a bound on the number of tasks contributing carry-in, since at time point $t_0 - 1$, there have to be strictly less than M higher priority tasks active. We state that property as the following lemma:

Lemma 4.1. *There are at most $M - 1$ tasks having carry-in, and for each task τ_i, the carry-in is at most $C_i - 1$.*

Proof. By discussion above, similar to [53].

Comparing with the original level-k busy period, with which one has to assume that all higher-priority tasks have carry-in, the over-estimation of the interference has been significantly reduced.

However, as introduced here, this busy period extension technique is not for free, since it is generally unknown when t_0 actually is, i.e., φ is an unknown variable. To solve this, [53] derives an upper bound (denoted by Φ^B) of all φ's values that need to be checked. The schedulability test is then conducted by enumerating every value of φ in $[0, \Phi^B]$. Finally, J_k (and therefore τ_k) is determined to be schedulable if the test can succeed with every value in $[0, \Phi^B]$.

Fig. 4.1 Extended level-k busy period for constrained-deadline task sets

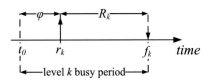

The upper bound Φ^B is pseudo-polynomial in the values of the task parameters, and in practice usually very large—especially for task sets with large parameter scales and/or with high utilizations. The RTA procedure itself (finding the fixed point) is also generally of pseudo-polynomial complexity, and requires quite a number of iterations in practice. Therefore, if one trivially conducts the RTA on each value of φ in $[0, \Phi^B]$, the complexity of the analysis would be very high in practice, and thus not practically usable. However, as we will see, the RTA procedure in our setting actually needs to be conducted only one single time with $\varphi = 0$, to get the same safe WCRT upper bound as enumerating all possible values of φ.

4.4.2   Workload and Interference

Before introducing the RTA procedure in detail, we will show bounds for workload and interference of tasks τ_i in a busy period of length x. These will later be used in the analysis.

4.4.2.1   Workload

The *workload* of a task in a certain busy period is the length of the accumulated execution time of that task within the busy period. We use $W_k(\tau_i, x)$ to denote an upper bound of the workload of each task τ_i with higher priority than the analyzed task τ_k in the level-k busy period of length x. From Lemma 4.1, we already know that there are at most $M - 1$ tasks doing carry-in, and all other tasks do not provide carry-in. So we define two types of workload:

- $W_k^{NC}(\tau_i, x)$ denotes the workload bound if τ_i does *not* have a carry-in job;
- $W_k^{CI}(\tau_i, x)$ is the workload bound if τ_i has a carry-in job.

To compute them, we have the following lemma.

Lemma 4.2. *The workload bounds can be computed with*

$$W_k^{NC}(\tau_i, x) = \left\lfloor \frac{x}{T_i} \right\rfloor \cdot C_i + [x \bmod T_i]^{C_i} \tag{4.3}$$

$$W_k^{CI}(\tau_i, x) = \left\lfloor \frac{[x - C_i]_0}{T_i} \right\rfloor \cdot C_i + C_i + \alpha \tag{4.4}$$

where $\alpha = [[x - C_i]_0 \bmod T_i - (T_i - R_i)]_0^{C_i-1}$.

Proof. Similar to reasoning in [53, 134], see Fig. 4.2.

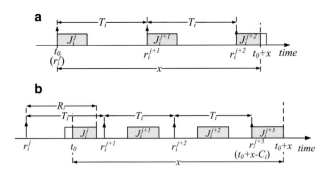

Fig. 4.2 Computing $W_k^{NC}(\tau_i, x)$ and $W_k^{CI}(\tau_i, x)$. (**a**) $W_k^{NC}(\tau_i, x)$. (**b**) $W_k^{CI}(\tau_i, x)$

Due to space limitations, we chose not to present a detailed proof and just show some intuition for the computation in Fig. 4.2. We would like to point out two important issues:

First, α in the computation of $W_k^{CI}(\tau_i, x)$ represents the carry-in of τ_i, which is limited to $C_i - 1$ according to Lemma 4.1. Further, the carry-in job is guaranteed to finish its computation within its response time R_i. Since we do the RTA for each task in their priority order, a bound of R_i is already known for each higher-priority task τ_i when computing $W_k^{CI}(\tau_i, x)$ for τ_k.

Second, and most important, we see from Eqs. (4.3) and (4.4) that both $W_k^{NC}(\tau_i, x)$ and $W_k^{CI}(\tau_i, x)$ are *independent of* φ. This means that, given only the length x of the level-k busy period, we always get the same result of $W_k^{NC}(\tau_i, x)$ and $W_k^{CI}(\tau_i, x)$, no matter when t_0 is (i.e., how large φ is). This key observation enables us to greatly reduce the computational efforts necessary to derive the safe WCRT bound.

4.4.2.2 Interference

As in [BC-RTA], we define the *interference* $I_k(\tau_i, x)$ of τ_i in the level-k busy period of length x. The interference denotes the part of the workload that can actually prevent τ_k from running. It can be less than the workload as we already discussed in Sect. 4.3.2. By carefully setting bounds, the analysis precision can be greatly improved.

Similar to the workload, we also use two types of $I_k(\tau_i, x)$: We use $I_k^{NC}(\tau_i, x)$ to denote the bound on the interference of τ_i to τ_k during a busy period of length x if τ_i does *not* have a carry-in job, while we use $I_k^{CI}(\tau_i, x)$ if τ_i has a carry-in job. Both values can be calculated with:

$$I_k^{NC}(\tau_i, x) = [W_k^{NC}(\tau_i, x)]_0^{x-C_k+1} \tag{4.5}$$

$$I_k^{\mathsf{CI}}(\tau_i, x) = [W_k^{\mathsf{CI}}(\tau_i, x)]_0^{x - C_k + 1} \tag{4.6}$$

As discussed in Sect. 4.3.2, the upper bound of the interference needs to be $x - C_k + 1$ rather than $x - C_k$.

We now define the *total interference* $\Omega_k(x)$, as the maximal value of the sum of all higher-priority tasks' interference among all possible cases with

$$\Omega_k(x) = \max_{(\tau^{\mathsf{NC}}, \tau^{\mathsf{CI}}) \in \mathscr{L}} \left(\sum_{\tau_i \in \tau^{\mathsf{NC}}} I_k^{\mathsf{NC}}(\tau_i, x) + \sum_{\tau_i \in \tau^{\mathsf{CI}}} I_k^{\mathsf{CI}}(\tau_i, x) \right), \tag{4.7}$$

where $\mathscr{L} \subseteq \tau \times \tau$ is the set of all partitions of the set $\tau_{<k} = \{\tau_1, \ldots, \tau_{k-1}\}$ into τ^{NC} and τ^{CI}, such that $\tau^{\mathsf{NC}} \cup \tau^{\mathsf{CI}} = \tau_{<k}$, $\tau^{\mathsf{NC}} \cap \tau^{\mathsf{CI}} = \emptyset$ and $|\tau^{\mathsf{CI}}| \leq M - 1$. By taking the maximum over this set, $\Omega_k(x)$ describes the maximal total interference when at most $M - 1$ are having carry-in, and all the others do not have carry-in. According to Lemma 4.1, $M - 1$ is the maximal number of tasks with carry-in, so indeed, $\Omega_k(x)$ expresses the maximal interference of higher-priority tasks to a task τ_k during a level-k busy period of length x.

Note that $\Omega_k(x)$ can be computed in linear time, since it is sufficient to find the $M - 1$ maximal values of the difference $I_k^{\mathsf{CI}}(\tau_i, x) - I_k^{\mathsf{NC}}(\tau_i, x)$, as pointed out in [53].

We state an important lemma about $\Omega_k(x)$.

Lemma 4.3. *For job J_k and all $x < f_k - t_0$, the following holds:*

$$\left\lfloor \frac{\Omega_k(x)}{M} \right\rfloor > x - C_k \tag{4.8}$$

Proof. The proof is given in Appendix 4.6.

Intuitively, the lemma states that, if we let the level-k busy period end before J_k's finish time, the total interference of all higher-priority tasks is large enough to prevent J_k from being finished within the level-k busy period. This in turn indicates that the level-k busy period actually has not reached its end, and thereby should continue to increase. Thus, this property of $\Omega_k(x)$ enables the iterative RTA procedure as will be presented in the next section.

4.4.3   The RTA Procedure

After defining an upper bound $\Omega_k(x)$ of the total interference to a task τ_k in a level-k busy period of length x, we present now how to use this for conducting the response time analysis for τ_k. The level-k busy period begins at the time point t_0, which is φ time units before r_k, the release of τ_k's job (see Sect. 4.4.1). In general, φ is an open variable. For a moment, we assume that the length φ of the busy period extension is given, and consider the particular $t_0 = r_k - \varphi$ derived from that. The following lemma expresses the response time analysis for such a particular t_0.

Lemma 4.4. *Given a $\varphi \geq 0$, let χ be the minimal solution of the recursive equation*

$$x = \left\lfloor \frac{\Omega_k(x)}{M} \right\rfloor + C_k. \tag{4.9}$$

Then $\chi - \varphi$ is an upper bound of τ_k's response time with this particular $t_0 = r_k - \varphi$.

Proof. Suppose the real worst-case response time of τ_k with $t_0 = r_k - \varphi$ is R, and assume $\chi - \varphi < R$ for the sake of contradiction. Since R is the real WCRT, there is a job sequence in which a job J_k exhibits this response time, i.e., $f_k - r_k = R$. It follows:

$$\chi < R + \varphi = f_k - r_k + \varphi = f_k - t_0$$

Thus, Lemma 4.3 applies, and therefore (4.8) holds with $x = \chi$. This contradicts the assumption of χ being a solution of (4.9). $\qquad\blacksquare$

Note that for all $k \leq M$, the minimal solution of (4.9) is trivially C_k, since in that case, $\Omega_k(x) < M$ for $x \leq C_k$ by definition. This matches the intuition that the M highest priority tasks don't suffer any interference, since M processors are available to accommodate them independently.

We have now seen how an upper bound of the response time of τ_k can be derived, if a particular φ is given. Since φ is an open variable, we need to find an upper bound of all response times for all φ to get a safe bound for the response time in general. Naively, it would seem that we have to enumerate all possible values of φ and solve (4.9), to obtain a safe upper bound of τ_k's WCRT.

However, as mentioned in Sect. 4.4.2, the computation of $W_k^{NC}(\tau_i, x)$ and $W_k^{CI}(\tau_i, x)$ is *independent of φ*. Thus, it turns out that $\Omega_k(x)$ is also independent of φ. Therefore, no matter what value of φ we are dealing with, the solution of Eq. (4.9) is always the same. And since $\chi - \varphi$ is an upper bound of τ_k's response time (with that particular φ), the maximal value and therefore general response time bound is χ. Thus, we only need to do the RTA according to Lemma 4.4 with $t_0 = r_k$ (i.e., $\varphi = 0$). The derived solution is guaranteed to be an upper bound of τ_k's WCRT.

Note that this observation can be regarded as a kind of *critical instant* for multiprocessor fixed-priority scheduling. In the context of our analysis, $\varphi = 0$ is guaranteed to be worst among all cases. Namely, we get the worst case when the earliest time instant after which all processors are occupied by higher-priority tasks occurs just before the release of τ_k. Still, since this does not provide precise information about the worst-case release times of the higher priority tasks, we are left with a *set* among which the real critical instant is found—but this set is significantly smaller than the whole space of possible job sequences.

We summarize the conclusion as the following theorem.

Theorem 4.1 ([OUR-RTA]). *Let* χ *be the minimal solution of the following Eq. (4.10) by doing an iterative fixed point search of the right-hand side starting with* $x = C_k$.

$$x = \left\lfloor \frac{\Omega_k(x)}{M} \right\rfloor + C_k \qquad (4.10)$$

Then χ *is an upper bound of* τ_k's *WCRT.*

Proof. By Lemma 4.4 and the above discussion.

Note that the iterative fixed point search should terminate with an "unschedulable" result as soon as $x > D_k$.

Since $\Omega_k(x)$ in [OUR-RTA] is no larger than $\sum_{i<k} I_k(\tau_i, x)$ in [BC-RTA], [OUR-RTA] dominates [BC-RTA] in the sense that an upper bound of the WCRT derived by [OUR-RTA] is guaranteed to be no larger than the one derived by [BC-RTA].

4.5   Performance Evaluation

We evaluate the performance of the proposed RTA technique in terms of acceptance ratio. We follow the method in [135] to generate task sets: A task set of $M + 1$ tasks is generated and tested. Then we iteratively increase the number of tasks by 1 to generate a new task set, and all the schedulability tests are run on the new task set. This process is iterated until the total processor utilization exceeds M. The whole procedure is then repeated, starting with a new task set of $M + 1$ tasks, until a reasonable sample space has been generated and tested. This method of generating random task sets produces a fairly uniform distribution of total utilizations, except at the extreme end of low utilization.

The default setting of the experiments whose results we show in Fig. 4.3 is as follows: the priorities are assigned according to global Deadline Monotonic scheduling; the number of processors is 6; for each task τ_i, T_i is uniformly distributed in $[10, 30]$. For each subfigure, the range of U_i and D_i/T_i is tuned (see the caption of each subfigure), to get task sets with different characteristics.

In all the figures, the curve "Sim" denotes the acceptance ratio of simulations. Since it is not computationally feasible to try all possible task release offsets and inter-release separations exhaustively in simulations, all task release offsets are set to be zero and all tasks are released periodically, and simulation is run for the hyper-period of all task periods. Simulation results obtained under this assumption may sometimes determine a task set to be schedulable even though it is not, but they can serve as a coarse upper bound of the ratio of all schedulable task sets.

Figure 4.3 shows the comparison between [OUR-RTA] and previous work for constrained-deadline task sets. In [129], [BC-RTA] has been compared with all state-of-the-art analysis for constrained-deadline task sets at that time, and shown clear

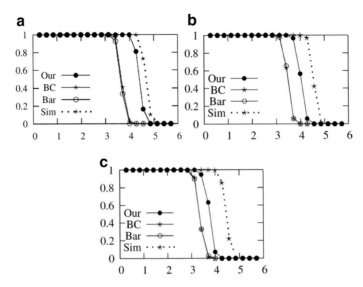

Fig. 4.3 Acceptance ratio: X-axis is total utilization $\sum_i U_i$; Y-axis is acceptance ratio. (**a**) $U_i \in [0.05, 0.1]$, $\frac{D_i}{T_i} \in [0.8, 1]$. (**b**) $U_i \in [0.05, 0.1]$, $\frac{D_i}{T_i} \in [0.8, 1]$. (**c**) $U_i \in [0.05, 0.1]$, $\frac{D_i}{T_i} \in [0.8, 1]$

performance improvement. Thus, we compare our analysis [OUR-RTA] (denoted by "Our") with [BC-RTA] (denoted by "BC") and the schedulability test from [53][1] (denoted by "Bar"), which is not included in the comparison in [129]. Another recent work by Bertogna et al. [136] is not included in Fig. 4.3 because it is outperformed by [BC-RTA]. It can be seen from the evaluation that [OUR-RTA] has non-trivial performance improvement over the others, especially with task sets with low utilizations.

4.6   Conclusions

We have developed a new technique to derive response time bounds for global fixed-priority scheduling on multiprocessors. This work made contributions in two folds: (1) The analysis precision has been significantly improved against previous work and, (2) To our best knowledge, this is the first work to study RTA of arbitrary-deadline systems on multiprocessors. We have used intensive experiments with randomly generated task sets to evaluate the performance of the proposed analysis techniques, in terms of both precision and efficiency. Experiments show that the proposed analysis has significant improvement of the analysis precision over existing methods, and can easily handle real-life-scale task systems. For future

---

[1]Works on global EDF scheduling, however, it can be easily adapted to global fixed-priority scheduling [53].

work, we will extend the proposed techniques to deal with platforms and task systems with shared resources and task synchronization.

Appendix: Proof of Lemma 4.3

Proof. We recall that by definition, $\Omega_k(x)$ consists of two sums over the sets τ^{NC} and τ^{CI} which are a partitioning of τ such that $\Omega_k(x)$ is maximal:

$$\Omega_k(x) = \sum_{\tau_i \in \tau^{\mathrm{CI}}} I_k^{\mathrm{CI}}(\tau_i, x) + \sum_{\tau_i \in \tau^{\mathrm{NC}}} I_k^{\mathrm{NC}}(\tau_i, x)$$

Let $\vartheta^{\mathrm{CI}} \subseteq \tau^{\mathrm{CI}}$ and $\vartheta^{\mathrm{NC}} \subseteq \tau^{\mathrm{NC}}$ be subsets of both partitions, such that

$$\forall \tau_i \in \vartheta^{\mathrm{CI}} : W_k^{\mathrm{CI}}(\tau_i, x) > x - C_k + 1$$

$$\forall \tau_i \in \vartheta^{\mathrm{NC}} : W_k^{\mathrm{NC}}(\tau_i, x) > x - C_k + 1$$

Thus, $\vartheta := \vartheta^{\mathrm{CI}} \cup \vartheta^{\mathrm{NC}}$ captures the relatively "dense" tasks of τ. Using this, $I_k^{\mathrm{CI}}(\tau_i, x, h)$ and $I_k^{\mathrm{NC}}(\tau_i, x, h)$ can be rewritten using $W_k^{\mathrm{CI}}(\tau_i, x)$ and $W_k^{\mathrm{NC}}(\tau_i, x)$ in the definition of $\Omega_k(x)$:

$$\Omega_k(x) = |\vartheta| \cdot (x - C_i + 1) + \sum_{\tau_i \in \tau^{\mathrm{CI}} \setminus \vartheta^{\mathrm{CI}}} W_k^{\mathrm{CI}}(\tau_i, x) + \sum_{\tau_i \in \tau^{\mathrm{NC}} \setminus \vartheta^{\mathrm{NC}}} W_k^{\mathrm{NC}}(\tau_i, x) \quad (4.11)$$

We consider the case of $|\vartheta| < M$. (Otherwise, the lemma obviously holds.)

Now let $x < f_k - t_0$, as in the assumption of the lemma, so the Job J_k is still active at time point x. Thus, only at strictly less than C_k time points of the interval $[t_0, t_0 + x)$, J_k was able to run. Now we know that all tasks from ϑ could keep at most $|\vartheta|$ processors busy at each time unit during the interval. It follows that the remaining tasks (those from $\tau \setminus \vartheta$) kept the remaining $M - |\vartheta|$ processors busy for at least $x - C_k + 1$ time units during the interval (otherwise, J_k would have been able to execute for C_k time units and thus finish until $t_0 + x$). Consequently, the tasks from $\tau \setminus \vartheta$ must have generated a workload of at least $(M - |\vartheta|) \cdot (x - C_k + 1)$ over the considered x time units. Since $W_k^{\mathrm{CI}}(\tau_i, x)$ and $W_k^{\mathrm{NC}}(\tau_i, x)$ are upper bounds of their workloads, we have

$$\sum_{\tau_i \in \tau^{\mathrm{CI}} \setminus \vartheta^{\mathrm{CI}}} W_k^{\mathrm{CI}}(\tau_i, x) + \sum_{\tau_i \in \tau^{\mathrm{NC}} \setminus \vartheta^{\mathrm{NC}}} W_k^{\mathrm{NC}}(\tau_i, x) \geq (M - |\vartheta|) \cdot (x - C_k + 1) \quad (4.12)$$

From (4.11) and (4.12) it follows

$$\Omega_k(x) \geq M \cdot (x - C_k + 1),$$

which is equivalent to the lemma.

Chapter 5
Analyzing Non-preemptive Global Scheduling

Non-preemptive scheduling is usually considered inferior to preemptive scheduling for time critical systems, because the non-preemptive block would lead to poor task responsiveness. Although this is true in single-processor scheduling, we found by empirical simulation experiments that it is not necessarily the case in multiprocessor scheduling. Additionally, non-preemptive scheduling enjoys other benefits like lower implementation complexity and run-time overhead. So non-preemptive scheduling may be a better alternative compared to preemptive scheduling for a considerable part of real-time applications on multiprocessor/multi-core platforms.

As the technical contribution, we study the schedulability analysis of global non-preemptive fixed-priority scheduling (NP-FP) on multiprocessors. We propose schedulability tests for NP-FP, building upon the "problem window analysis" by Baruah [53] for preemptive scheduling. We firstly derive a linear-time general schedulability test condition that works for not only NP-FP, but also any other work-conserving non-preemptive scheduling algorithm. Then we improve the analysis and present a test condition of quadratic time-complexity for NP-FP, which has significant performance improvement comparing to the first one. A notable advantage of our proposed test conditions is, while the test in [53] needs to enumerate for a large number of possible problem window sizes, our proposed test conditions only need to be conducted with a single problem window size, and thereby are significantly more efficient. Experiments with randomly generated task sets are conducted to evaluate the performance of the proposed test conditions.

5.1 Introduction

Non-preemptive scheduling has received considerable less attention in the research community compared to preemptive scheduling. However, non-preemptive scheduling is widely used in industry practice, and it may be preferable to preemptive

© Springer International Publishing Switzerland 2016
N. Guan, *Techniques for Building Timing-Predictable Embedded Systems*,
DOI 10.1007/978-3-319-27198-9_5

scheduling for a number of reasons [137]: non-preemptive scheduling algorithms
are easier to implement and have lower run-time overhead than preemptive scheduling algorithms; the overhead of preemptive scheduling algorithms is more difficult
to characterize and predict than that of non-preemptive scheduling algorithms due
to inter-task interference caused by caching and pipelining. These benefits of non-
preemptive scheduling are more important on multiprocessor platforms, where the
task migration overhead is even higher and more difficult to predict. However,
this problem is much less severe for non-preemptive scheduling, where each task
instance runs to completion on one processor, and task migrations only happen at
task instance boundaries.

Multiprocessor systems are becoming more and more important, with industry trends such as multi-core processors and Multiprocessor Systems-on-a-Chip
(MPSoC). Therefore, real-time scheduling and schedulability analysis for multiprocessor systems have become an important research area. Multiprocessor scheduling
algorithms can be classified into two categories: *partitioned scheduling*, where
each task is assigned to a processor, and task migration between processors is not
allowed; and *global scheduling*, where each task can migrate between different
processors either during execution of one of its instances or between different
instances. The analysis of global scheduling is significantly more difficult than
partitioned scheduling [53]. In this chapter we focus on global scheduling.

Non-preemptive scheduling is considered inferior for real-time systems because
of its poor responsiveness. In a single processor system, a high-priority task may be
blocked by a low-priority task for a long time due to non-preemptiveness, and thus
miss its deadline. However, this problem is less severe in a multiprocessor system,
since the natural parallelism of the multiprocessor platform can mitigate the harmful
effect of non-preemptive blocking. Even if several processors are occupied by low-
priority tasks with large execution time, high-priority tasks still have chances to
execute on other processors and meet their deadlines. We have conducted simulation
experiments to compare the performance, measured in terms of the percentage
of task sets that can be feasibly scheduled with a given scheduling algorithm,
of global preemptive fixed-priority scheduling (P-FP) and non-preemptive fixed-
priority scheduling (NP-FP). To our surprise, under many parameter settings
NP-FP actually outperforms P-FP (the simulation experiments will be presented
in detail in Sect. 5.8). Note that we did not count context switch overheads in
the simulation experiments, which would reduce the performance of preemptive
scheduling algorithms further compared to non-preemptive scheduling algorithms.
This leads us to believe that non-preemptive scheduling may be a better alternative
compared to preemptive scheduling with respect to real-time performance for
certain applications on multiprocessor platforms.

In this chapter, we address the schedulability analysis problem of NP-FP for
sporadic task sets on identical multiprocessors. The proposed analysis techniques
are built upon Baker's "problem window" analysis framework [52] and Baruah's
technique of "problem window extension" to bound the number of tasks doing
carry-in [53]. To the best of our knowledge, this is the first work to study the
schedulability analysis problem of NP-FP. A notable advantage of our proposed test

conditions is, while all the previous works with the "problem window extension" [53, 138] are of pseudo-polynomial complexity and need to enumerate for a large number of possible sizes of the problem window, our proposed test conditions only need to be conducted for a single problem window size, and thereby are significantly more efficient.

We first present a general schedulability test condition of linear computational complexity, which works on any work-conserving non-preemptive multiprocessor scheduling algorithm (therefore, it can be used for NP-FP). Then we refine the "problem window" and proposed the second schedulability test condition specially found for NP-FP, which is of quadratic computational complexity. Experiments with randomly generated task sets are conducted to evaluate the performance of the proposed test conditions.

5.2 Related Work

5.2.1 Preemptive Scheduling

All scheduling algorithms discussed in this paragraph are implicitly "global preemptive," e.g., we refer to "global preemptive EDF" as "EDF" for short. Goossens et al. [139] introduced a schedulability test with polynomial time-complexity for periodic task sets scheduled by EDF based on the resource-augmentation technique [140]. Similar techniques were used in [50] to derive schedulability tests for tasks with limited utilization scheduled by Rate Monotonic (RM) scheduling. Baker [52] presented schedulability tests for both EDF and Deadline Monotonic (DM) scheduling by determining the necessary conditions on the parameters of all the tasks to cause a given task τ_k's instance to miss its deadline. Based on Baker's idea, Bertogna et al. [54] observed that the work done in parallel with a task instance does not need to be added to its interference, and provided a new test condition with polynomial time-complexity that can sometimes outperform Baker's test condition. Baruah [53] improved Baker's approach by "problem window extension" to reduce the over-estimation of the so-called carry-in jobs, and provided a test condition with pseudo-polynomial time-complexity. It has higher acceptance ratio than previous test conditions for task systems that satisfy the following conditions: the number of tasks n is significantly larger than the number of processors m (i.e., $n \gg m$), or the parameters of different tasks have widely varying orders of magnitude [53]. Andersson et al. [56] first used the approximate response time analysis for multiprocessor scheduling, which was later improved by Bertogna et al. [129] by applying their techniques in [54] and exploring task slack time to reduce the degree of pessimism in the computation of the approximate response time. Bertogna et al. [136] also applied the same idea of slack time exploration to schedulability tests for EDF and FP scheduling. Recently, Guan et al. [55] applied Baruah's idea of problem window extension to the response time analysis of FP scheduling, and proposed new response time bounds for FP scheduling that dominates the result in [129].

5.2.2   Non-preemptive Scheduling

For single-processor scheduling, Jeffay et al. [137] considered non-preemptive algorithms for scheduling periodic or sporadic task systems with relative deadlines equal to periods under the work-conserving assumption and presented an exact schedulability test of pseudo-polynomial time-complexity for a periodic or sporadic task set under non-preemptive EDF scheduling on a single processor. George et al. [37] addressed general task models in which relative deadlines and periods are not necessarily related, and established exact schedulability tests for both non-preemptive EDF and non-preemptive fixed-priority scheduling on a single processor with pseudo-polynomial time complexity. Baruah et al. [141] addressed schedulability analysis for non-preemptive recurring tasks, which is the general form of non-preemptive sporadic tasks, and showed that the non-preemptive schedulability analysis problem can be reduced to a polynomial number of preemptive schedulability analysis problems.

For multiprocessor scheduling, Baruah [142] proposed a sufficient but not necessary polynomial-time schedulability test condition [TEST-BAR] for global non-preemptive scheduling for periodic task sets, which can be easily generalized to sporadic task sets. [TEST-BAR] used a technique similar to [139] and took into account the extra interference time caused by non-preemption. According to [TEST-BAR], a task set τ is NP-EDF schedulable on m processors if

$$V_{sum}(\tau) \leq m - (m-1)V_{max}(\tau) \tag{5.1}$$

where

$$V_{sum}(\tau) = \sum_{\tau_i \in \tau} V_i, \quad V_{max}(\tau) = \max_{\tau_i \in \tau} V_i$$

$$V_i = \begin{cases} \dfrac{C_i}{D_i - C_{max}} & D_i > C_{max} \\ \infty & D_i \leq C_{max} \end{cases}$$

and C_{max} is the maximum execution time among all tasks. It is obvious that a task set with arbitrarily low utilization cannot pass the test if $C_{max} \geq D_{min}$, where D_{min} denotes the minimal D_i among all tasks. Intuitively, it means that for any task instance J_k, if there is some task with an execution time large enough to cover its relative deadline D_k, J_k will definitely miss its deadline. This is true for single-processor scheduling, but not necessarily true for multiprocessor scheduling, since even if some processors are occupied for a long time by a task instance with a large C_i, other task instances can execute on different processors to meet their deadlines. Recently Baruah's idea of problem window extension [53] has been applied to non-preemptive multiprocessor scheduling [134, 138]. However, both [134, 138] focused on EDF scheduling while in this paper we focus on Fixed-Priority scheduling.

Moreover, the test conditions proposed in this chapter are of polynomial (linear and quadratic respectively) computational complexity and applicable to the dense time model, while all previous schedulability tests applying the "problem window extension" [53, 134, 138] are of pseudo-polynomial computational complexity and only applicable to the discrete time model.

5.3 System Model and Notations

We assume that a multiprocessor platform consists of M identical processors. A sporadic task set τ consists of N sporadic tasks. A sporadic task is denoted by $\tau_i = (C_i, D_i, T_i)$, where C_i is the worst-case execution time, D_i is the relative deadline, and T_i is the minimum inter-release separation, also referred to as its period. For each task τ_i we assume $D_i \leq T_i$ and define $S_i = D_i - C_i$. The *utilization* of task τ_i is defined as $U_i = \frac{C_i}{T_i}$, and we use $U(\tau)$ to denote the sum of U_i of all $\tau_i \in \tau$. Each task has a unique priority. We use $hp(k)$ to denote the set of tasks with *higher* priorities than τ_k, and $hep(k) = hp(k) \cup \{\tau_k\}$ the set of tasks whose priorities are *no lower* than τ_k. Similarly, $\mathcal{L}(k)$ is the set of tasks with *lower* priorities than τ_k and $lep(k) = \mathcal{L}(k) \cup \{\tau_k\}$ the set of tasks whose priorities are *no higher* than τ_k.

Such a sporadic task τ_i generates a potentially infinite sequence of *jobs* (also called *task instances*) with successive releases separated by at least T_i time units. We use J_i^p to denote the pth job of τ_i. We also use J_i to denote a job of τ_i in general if we do not want to specify which job it is. Each job J_i has a release time (arrival time) r_i and an absolute deadline $d_i = r_i + D_i$. We use $l_i = d_i - C_i$ to denote J_i's *latest feasible start time* which is the latest time point for J_i to start execution in order to meet its deadline.

A job is *pending* at time instant t if it was released before t and has uncompleted work at t. A task is *pending* at time instant t if it has a pending job at t.

We use $\Theta(t, \tau_k)$ to denote the set of pending tasks with higher priorities than τ_k at time t, i.e.,

$$\Theta(t, \tau_k) = \{\tau_i | \tau_i \in hp(\tau_k) \wedge \tau_i \text{ is pending at } t\} \tag{5.2}$$

$|\Theta(t, \tau_k)|$ denotes the number of elements in $\Theta(t, \tau_k)$.

For simplicity of expression, we further use the following notations to express that a value A is "limited" if it is below or above a certain threshold B or C, respectively: $[A]_B = \max(A, B), [A]^C = \min(A, C)$, and $[A]_B^C = [[A]_B]^C$. This expression just keeps the value A if it is within the interval $[B, C]$, otherwise it is B if $A < B$ or C if $A > C$.

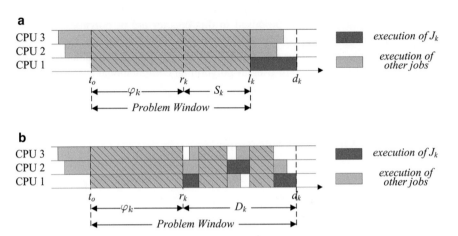

Fig. 5.1 Problem window in (**b**) preemptive and (**a**) non-preemptive scheduling

5.4   The General Schedulability Test

In this section we present a general schedulability test condition that works for any work-conserving non-preemptive scheduling algorithm.

Given a work-conserving non-preemptive scheduling algorithm Schd, and suppose a task set τ is non-schedulable by Schd, let J_k be the first job missing its deadline, and r_k is the release time of J_k. Let t_o denote the earliest time instant before r_k, such that there are M tasks running at any time instant $t \in [t_o, r_k]$, and let $\varphi_k = r_k - t_o$, as shown in Fig. 5.1a. Since all processors are idle when the system starts, there always exists such a t_o.

By the definition of t_o, we know that all processors are continuously busy during $[t_o, r_k]$. Since preemption is not allowed, once a job starts execution, it will run to completion without interruption. So if J_k starts to execute before its latest start time l_k, it will be able to finish execution before the deadline d_k. Since Schd is work-conserving, we know that *in order for J_k to miss its deadline, all M processors must be continuously busy in the time interval $[t_o, l_k]$*. What happens after l_k has no effect on the schedulability of J_k. We name the time interval $[t_o, l_k]$ as *problem window*, as shown in Fig. 5.1a.

The definition of the problem window above follows the idea in [53] for preemptive scheduling but is slightly different. With the problem window in [53] all M processors are continuously busy in the time interval $[t_o, r_k]$, but do not have to be continuously busy in the time interval $[r_k, d_k]$ as long as the sum of the busy segments (shadow area in Fig. 5.1b) is large enough to cause τ_k to miss its deadline.

A necessary condition for the deadline miss to occur is that the worst-case workload in the problem window by all other jobs in the task set τ except J_k, is no less than $(\varphi_k + S_k)M$ (the shadow area in Fig. 5.1a). Since the critical instant is generally unknown in global multiprocessor scheduling, it is not possible to find the worst-case situation without exhaustively simulating the system. So we will compute the worst-case workload by each task in the problem window, and use the

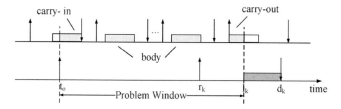

Fig. 5.2 Body, carry-in and carry-out of a task in the problem window $[t_o, l_k]$

sum of each task's workload as an upper bound of the overall worst-case workload in the problem window. The workload of a task τ_i in the problem window can be categorized into three parts:

- *carry-in*: the contribution of at most one job (called carry-in job) with release time earlier than t_o and deadline in the problem window;
- *carry-out*: the contribution of the last job (called carry-out job) with release time in the problem window and deadline outside the problem window;
- *body*: the contribution of the jobs except the carry-in job and carry-out job.

We always consider a carry-in job is executed as late as possible, a carry-out job is executed as early as possible and all jobs of a task are released periodically, as shown in Fig. 5.2. This is a pessimistic but safe approximation to count the workload of a task in the problem window [52].

By the definition of t_o, we have the following lemma:

Lemma 5.1. *At most $M - 1$ tasks have carry-in jobs.*

Proof. Recall that t_o is the earliest time instant before r_k, such that all M processors are busy at any $t \in [t_o, r_k]$. We use t_o^- to denote a time instant that is earlier than and arbitrarily close to t_o, so there are at most $M - 1$ tasks executing at t_o^-, and there is no other pending task at t_o^-. Since only the tasks pending at t_o^- have carry-in jobs, we know there are at most $M - 1$ tasks having carry-in jobs.

We use W to denote the total workload of the task set τ in the problem window, and we get an upper bound for W^1:

Lemma 5.2. *We sort all C_i in a non-increasing list, and use C_{M-1}^{sum} to denote the sum of the first $(M - 1)$ elements in this list. Then we have*

$$\mathsf{W} \leq C_{M-1}^{\text{sum}} + \sum_{\tau_i \in \tau} \left\lfloor \frac{\varphi_k + S_k}{T_i} \right\rfloor C_i + \sum_{\tau_i \in \tau} C_i \qquad (5.3)$$

---

[1] One can get a tighter bound on W with a more precise calculation of carry-in/carry-out. However, to get a linear over-approximation of workload in the derivation of the first test condition in the following, a bound needs to be relaxed anyway, which leads to the same result as using this simple bound.

Proof. For each task $\tau_i \in \tau$, its workload in the problem window consists of three parts: carry-in, body, and carry-out. The carry-in of each job is bounded by its computation time, and by Lemma 5.1 we also know at most $M - 1$ tasks have carry-in jobs, so the total work by all carry-in jobs is bounded by $C_{\text{M-1}}^{\text{sum}}$ [the first item in the RHS of Inequality (5.3)]. The number of body jobs of task τ_i is bounded by $\lfloor \frac{\varphi_k + S_k}{T_i} \rfloor$, so the second item in the RHS of Inequality (5.3) is an upper bound of the workload of all body jobs. The carry-out of each job is also bounded by its computation time, so the second item in the RHS is an upper bound of the workload by all carry-out jobs.

The main idea of the proposed schedulability test condition is as follows: If one can guarantee that this upper bound of total workload is smaller than the total processor capacity in the problem window, i.e.,

$$C_{\text{M-1}}^{\text{sum}} + \sum_{\tau_i \in \tau} \left\lfloor \frac{\varphi_k + S_k}{T_i} \right\rfloor C_i + \sum_{\tau_i \in \tau} C_i < (\varphi_k + S_k)M \qquad (5.4)$$

it can be concluded that the workload of the task set is not enough to make all processors continuously busy during the problem window, so J_k can start execution no later than l_k. This contradicts with our assumption that J_k misses its deadline, so it implies J_k can meet its deadline. By applying this reasoning to each task in τ, we can get a sufficient schedulability condition for Schd. Note that a difficulty to derive the needed condition is that the unknown variable φ_k appears on both sides of Inequality (5.4). However, we observe that as φ_k increases, the proportion of the carry-in and carry-out in the overall workload of a task in the problem window tends to decrease, which means that the effect of the over-estimation of the carry-in and carry-out is more severe with smaller $\varphi_k + S_k$ values, with the extreme case of $\varphi_k = 0$. Thus we derive our first test condition for an arbitrary work-conserving non-preemptive scheduling algorithm Schd:

Theorem 5.1. [TEST-1] *A task set τ is schedulable by a work-conserving non-preemptive scheduling algorithm* Schd *on M processors if:*

$$U(\tau) < M - \frac{\sum_{\tau_i \in \tau} C_i + C_{\text{M-1}}^{\text{sum}}}{S_{min}} \qquad (5.5)$$

where S_{min} is the minimal S_i among all tasks.

Proof. We prove the theorem by contradiction. Assume a task set τ satisfies Inequality (5.5) but is non-schedulable, and τ_k is the first task missing its deadline. We know that the total workload W should not be smaller than $(\varphi_k + S_k)M$, and by Lemma 5.2 we have

$$C_{\text{M-1}}^{\text{sum}} + \sum_{\tau_i \in \tau} \left\lfloor \frac{\varphi_k + S_k}{T_i} \right\rfloor C_i + \sum_{\tau_i \in \tau} C_i \geq (\varphi_k + S_k)M$$

Since $\lfloor \frac{\varphi_k + S_k}{T_i} \rfloor C_i \leq (\varphi_k + S_k)\frac{C_i}{T_i}$ and $U(\tau) = \sum_{\tau_i \in \tau} \frac{C_i}{T_i}$, we have

$$C_{\text{M-1}}^{\text{sum}} + (\varphi_k + S_k)U(\tau) + \sum_{\tau_i \in \tau} C_i \geq (\varphi_k + S_k)M$$

$$\Rightarrow \sum_{\tau_i \in \tau} C_i + C_{\text{M-1}}^{\text{sum}} >\geq (\varphi_k + S_k)(M - U(\tau))$$

and since $\varphi_k \geq 0$ and $S_k \geq S_{min}$, we get

$$U(\tau) \geq M - \frac{\sum_{\tau_i \in \tau} C_i + C_{\text{M-1}}^{\text{sum}}}{S_{min}}$$

which contradicts our assumption that τ satisfies Inequality (5.5).

[TEST-1] works on any work-conserving non-preemptive scheduling algorithm, since no scheduling algorithm specific properties are used in its proof, except that the algorithm is work-conserving.

Note that [TEST-1] does not suffer from the disadvantage in [TEST-BAR] that any task set with $C_{max} \geq D_{min}$ will be rejected.

U_i, $\sum_{i=1}^{N} C_i$ and S_{min} can all be computed in linear time, and we can use linear-time selection [143] to compute $C_{\text{M-1}}^{\text{sum}}$, so [TEST-1] has linear-time complexity.

5.5 The Improved Schedulability Test for NP-FP

In the last section we derived a sufficient schedulability test condition [TEST-1] that works on any work-conserving non-preemptive multiprocessor scheduling algorithm. [TEST-1] is safe, but very pessimistic, since it uses a grossly coarse upper bound on the task set's workload in the problem window. In this section, we will present a less pessimistic test condition for NP-FP by refining the definition of the "problem window" and deriving more precise bounds on the workload.

Again, we suppose a task set τ is non-schedulable by NP-FP, and let J_k be the first job missing deadline, and r_k is the release time of J_k. We will re-define the problem window. First, we adopt the definition of t_o in [138] to NP-FP: Let t_o be the earliest time instant before r_k, such that $\forall t \in [t_o, r_k)$ one of the following two properties holds:

1. $|\Theta(t, \tau_k)| = M$ and all tasks in $\Theta(t, \tau_k)$ are executing.
2. There are some tasks in $\Theta(t, \tau_k)$ not executing at t.

Recall that $\Theta(t, \tau_k)$ is the set of all pending tasks with *higher* priority than τ_k at t. If there does not exist such a t_o, we let $t_o = r_k$.

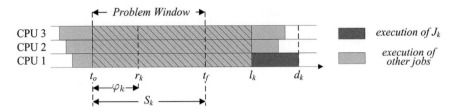

Fig. 5.3 The new problem window

We re-define the end of the problem window to be $t_f = t_o + S_k$, and the time interval $[t_o, t_f]$ is the new *problem window*, as illustrated in Fig. 5.3. Note that the length of the problem window is a fixed value S_k, although the start point (and thereby the end point) of the problem window is not fixed.

We have the following properties with the new problem window:

Lemma 5.3. *During the problem window $[t_o, t_f]$ all processors are occupied by jobs of tasks in hp(k) or jobs starting execution earlier than t_o.*

Proof. We will prove the lemma in two steps: First we prove during $[t_o, t_f]$ all processors are continuously busy; second we prove that any job executing during $[t_o, t_f]$ either is of a task in $hp(k)$ or has started execution before t_o.

Now we prove the first step. Remind the two conditions in the definition of t_o. For any time instant $t \in [t_o, r_k)$, all processors are busy if the first condition holds. If the second condition holds, some pending task cannot execute. And since NP-FP is work-conserving, we know all processors are busy at t, and thus all processors are continuously busy during $[t_o, r_k)$. Since J_k misses its deadline, we also know all processors are continuously busy during $[r_k, l_k]$. By the definition of t_f we have $t_f \leq l_k$, so we know that during $[t_o, t_f]$ all processors are continuously busy.

Next we prove the second step by contradiction. Suppose there is a job J_i of a task τ_i in $lep(k)$ and starting execution at t ($t \in [t_o, t_f]$). We distinguish two cases: (1) $t \in [t_o, r_k)$; (2) $t \in [r_k, t_f]$.

First we consider case (1). Since J_i starts execution at t, we know all tasks in $\Theta(t, \tau_i)$ are executing at t and $|\Theta(t, \tau_i)| < M$. Since τ_i's priority is no higher than τ_k, we have $\Theta(t, \tau_k) \subseteq \Theta(t, \tau_i)$. So we know that all tasks in $\Theta(t, \tau_i)$ are executing at t and $|\Theta(t, \tau_i)| < M$, which contradicts with the definition of t_o.

Then we consider case (2). Since J_k misses its deadline, no job with lower priority than τ_k can start execution in $[r_k, l_k]$. J_k's precedent job also cannot start executing in $[r_k, l_k]$ since each task's relative deadline is no larger than its period. So we know if J_i is of a task τ_i in $lep(k)$, it cannot start execution in $[r_k, l_k]$, and since $t_f \leq l_k$, it cannot start execution in $[r_k, t_f]$, which contradicts with our assumption about J_i.

So both cases lead to contradictions, by which we know any job executing during $[t_o, t_f]$ either is of a task in $hp(k)$ or starts execution earlier than t_o.

Lemma 5.4. *At most M tasks have carry-in jobs.*

Proof. Still we use t_o^- to denote a time instant that is earlier than t_o and arbitrarily close to t_o, so by the definition of t_o we know $|\Theta(t_o^-, \tau_k)| \neq M$ and all tasks in $\Theta(t_o^-, \tau_k)$ are executing. Since it is not possible for more than M tasks executing at the same time, the only possible case is $\Theta(t_o^-, \tau_k) < M$ and all tasks in $\Theta(t_o^-, \tau_k)$ are executing. A task in $hp(k)(\tau)$ has carry-in only if it's in $\Theta(t_o^-, \tau_k)$, so we know there are $|\Theta(t_o^-, \tau_k)|$ tasks in $hp(k)(\tau)$ having carry-in. A task in $\mathfrak{L}(k)(\tau)$ has carry-in only if it is executing at t_o^-. The number of tasks in $\mathfrak{L}(k)(\tau)$ executing at t_o^- is at most $M - |\Theta(t_o^-, \tau_k)|$, since $\Theta(t_o^-, \tau_k)$ processors have been occupied by tasks in $hp(k)(\tau)$ as mentioned above. Therefore, there are at most $|\Theta(t_o^-, \tau_k)| + (M - |\Theta(t_o^-, \tau_k)|) = M$ tasks doing carry-in.

We use $I_k^{NC}(\tau_i)$ to denote an upper bound of τ_i's workload in the problem window if τ_i has no carry-in job, and use $I_k^{CI}(\tau_i)$ to denote an upper bound of τ_i's workload if τ_i has a carry-in job. In the following we will show how to compute $I_k^{CI}(\tau_i)$ and $I_k^{NC}(\tau_i)$ by identifying the worst-case scenario of each task's workload in the problem window under NP-FP.

- Computing $I_k^{NC}(\tau_i)$

  - $\tau_i \in hp(k)$. The worst-case workload of task τ_i in this case is shown in Fig. 5.4 (the reasoning is rather simple so we omit it), according to which we can compute $I_k^{NC}(\tau_i)$ by:

$$I_k^{NC}(\tau_i) = \left\lfloor \frac{S_k}{T_i} \right\rfloor C_i + [S_k \bmod T_i]^{C_i} \tag{5.6}$$

  - $\tau_i \in lep(k)$. By Lemma 5.3, we know that for any task $\tau_i \in lep(k)$, only the job starting execution before t_o can execute in the problem window, so τ_i's workload in the problem window is 0 if it has no carry-in job, i.e.,

$$I_k^{NC}(\tau_i) = 0 \tag{5.7}$$

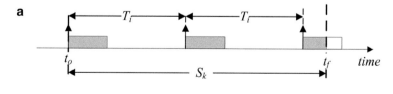

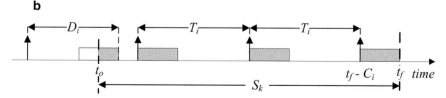

Fig. 5.4 The worst-case scenario of (a) $I_k^{NC}(\tau_i)$ and (b) $I_k^{CI}(\tau_i)$

- Computing $I_k^{CI}(\tau_i)$

  - $\tau_i \in hp(k)$. The worst-case workload of task τ_i in this case is shown in Fig. 5.4, by which we can compute $I_k^{CI}(\tau_i)$ by

$$I_k^{CI}(\tau_i) = [\lfloor [S_k - C_i]_0/T_i \rfloor C_i + C_i + \alpha]^{S_k} \qquad (5.8)$$

  where $\alpha = [[S_k - C_i]_0 \bmod T_i - (T_i - D_i)]_0^{C_i}$. Note that we pose an upper bound S_k on $I_k^{NC}(\tau_i)$, since the work actually done by a task in a time interval is no larger than the length of the problem window, and the part of its workload beyond the length of the problem window will actually not interfere with τ_k.
  - $\tau_i \in lep(k)$. By Lemma 5.3 we know that for any task $\tau_i \in lep(k)$, only its job starting execution before t_o can execute in the problem window, and we know there is at most one such job of τ_i, so we can compute $I_k^{CI}(\tau_i)$ by

$$I_k^{CI}(\tau_i) = [C_i]^{S_k} \qquad (5.9)$$

An important observation on the above computation is, $I_k^{NC}(\tau_i)$ and $I_k^{CI}(\tau_i)$ are completely independent on when is t_o. This is because the problem window defined in this section is an interval with fixed length S_k (Recall that the problem window in last section is defined as $[t_o, d_k]$. Since t_o is an unknown time point and d_k is a fixed time point, the length of the problem window is not fixed.).

Now we define Ω_k as the maximal value of the sum of all tasks's work in the problem window among all possible cases with

$$\Omega_k = \max_{(\tau^{NC}, \tau^{CI}) \in \mathscr{L}} \left(\sum_{\tau_i \in \tau^{NC}} I_k^{NC}(\tau_i) + \sum_{\tau_i \in \tau^{CI}} I_k^{CI}(\tau_i) \right), \qquad (5.10)$$

where $\mathscr{L} \subseteq \tau \times \tau$ is the set of all partitions of the task set τ into τ^{NC} and τ^{CI}, such that $\tau^{NC} \cup \tau^{CI} = \tau$, $\tau^{NC} \cap \tau^{CI} = \emptyset$, and $|\tau^{CI}| \leq M$. By taking the maximum over this set, Ω_k describes the maximal total interference when at most M tasks are having carry-in, and all the others do not have carry-in. According to Lemma 5.4, M is the maximal number of tasks with carry-in, so indeed, Ω_k expresses the maximal workload of the task set in the problem window. Note that Ω_k can be computed in linear time, since it is sufficient to find the M maximal values of the difference $I_k^{CI}(\tau_i) - I_k^{NC}(\tau_i)$, as pointed out in [53].

If one can show that Ω_k is smaller than the total processor capacity during the problem window, then τ_k can start execution in the problem window, i.e., no later than t_f, and since the end of the problem window t_f is no later than l_k, τ_k can meet its deadline. By applying this to each task in the task set, we get obtain the improved schedulability test condition for NP-FP:

Theorem 5.2. *The task set τ is NP-FP schedulable if $\forall \tau_k \in \tau$ we have*

$$\Omega_k < S_k M \qquad (5.11)$$

Both $I_k^{NC}(\tau_i)$ and $I_k^{CI}(\tau_i)$ can be computed in linear time, and we can use linear-time selection [143] to find the M maximal values of the difference $I_k^{CI}(\tau_i) - I_k^{NC}(\tau_i)$, so for each task $\tau_k \in \tau$ the test condition is of linear computational complexity, and the overall scheduling test is of quadratic computational complexity.

5.6 Sustainability

Baruah and Burns [144] introduced the concept of *sustainability* (also named as robustness [145] and predictability [146] in earlier literatures) for a schedulability test. A schedulability test is sustainable if a system that is determined to be schedulable under its worst-case specification remains schedulable when its actual run-time behavior is "better" than worst-case, e.g., when inter-release separations are increased, relative deadlines are increased or task execution times are reduced. The sustainability property is important. If a schedulability test is sustainable, then the system designer only needs to consider the worst-case parameter values to determine if the system is schedulable, instead of considering every possible execution time value in the interval (BCET, WCET) and/or the infinitely-many possible inter-release separation and deadline values in $[D_i, \infty)$ and $[T_i, \infty)$.

Recently Baker and Baruah [147] introduced a stronger concept *self-sustainability*. A schedulability test is self-sustainable if a system that is determined to be schedulable under its worst-case specification can still pass this schedulability test with "better" parameters. The property of self-sustainability is useful in incremental, interactive design processes. If a schedulability test is self-sustainable, it is guaranteed that relaxing timing constraints will not make a schedulable system (subsystem) unverifiable, thus one can safely relax the task parameters of the schedulable system in order to explore the design state space. It is easy to see that any self-sustainable schedulability test is also sustainable.

Now we present the (self-)sustainability properties of the proposed test conditions.

Theorem 5.3. *The schedulability test* [TEST-1] *is self-sustainable with regard to execution time, deadline, and inter-release separation.*

Proof. The proof follows directly from examination of Condition (5.5). □

Theorem 5.4. *The schedulability test* [TEST-2] *is sustainable with regard to execution time, deadline, and inter-release separation, and is self-sustainable with regard to inter-release separation.*

Proof. Let τ_k be a task that can pass the schedulability test [TEST-2]. [TEST-2] is independent of T_k, and both $I_k^{NC}(\tau_i)$ and $I_k^{CI}(\tau_i)$ are non-increasing with respect to T_i, so [TEST-2] still holds if we increase the inter-release separation of any task. Therefore [TEST-2] is self-sustainable with regard to inter-release separation, which also implies [TEST-2] is sustainable with regard to inter-release separation.

Next we show [TEST-2] is sustainable with regard to execution time. Still let τ_k be a task that can pass the schedulability test [TEST-2], which means any job J_k^p of τ_k can start execution no later than its latest start time l_k^p, so decreasing C_k will not make J_k^p missing its deadline. Now we consider relaxing the parameter of some other τ_i. Since both $I_k^{NC}(\tau_i)$ and $I_k^{CI}(\tau_i)$ are non-decreasing with respect to C_i, if we decrease τ_i's execution time C_i, τ_k can still pass [TEST-2], so τ_k is still schedulable.

At last we show [TEST-2] is sustainable with regard to relative deadline. In fixed-priority scheduling, the scheduling decision only depends on the tasks' priorities, but independent tasks' deadlines, which means for any concrete scheduling sequence, if we increase some task's relative deadline, the scheduling sequence is unchanged, by which we know a schedulable task τ_k will not become unschedulable if some task's relative deadline is increased.

Note that [TEST-2] is *not* self-sustainable with regard to execution times and relative deadlines. This happens when we decrease the execution time or increase the relative deadline of the analyzed task: Both decreasing the execution time and increasing the relative deadline will lead to a larger problem window size S_k. However, Ω_k is a piecewise linear function with respect to S_k and increasing S_k may lead to a greater increase on the LHS of [TEST-2] than its RHS. Consider the task set in Table 5.1 with $M = 2$, in which τ_5 is the task under analysis. With the current parameters $S_5 = 10$ and we have

$$\Omega_5 = 10 + 8 + 0.9 + 0.9 = 19.8 < 10 \times 2 = S_5 \times M \qquad (5.12)$$

(τ_1 and τ_2 have carry-in, while τ_3 and τ_4 do not), by which τ_5 is determined to be schedulable. However, if we decrease C_5 or increase D_5 by 1, then $S_5 = 11$ and we have

$$\Omega_5 = 11 + 8 + 1.8 + 1.8 = 22.6 > 11 \times 2 = S_5 \times M \qquad (5.13)$$

by which τ_5 is determined to be non-schedulable. However, decreasing execution time and increasing relative deadline of other tasks does not cause the analyzed task to violate the test condition [TEST-2]. So in the design phase, when a task's parameter becomes "better," the designer only needs to re-examine if this particular task still satisfies the test condition, but does not need to worry about other tasks.

Table 5.1 An example to show [TEST-2] is not self-sustainable with respect to C_k and D_k

Task	T_i	D_i	C_i
τ_1	10	10	6
τ_2	10	10	4
τ_3	10	10	0.9
τ_4	10	10	0.9
τ_5 (analyzed task)	14	14	4

5.7   Performance Evaluation

In this section, we use synthetic task sets to compare the performance of these schedulability test conditions in terms of acceptance ratio, i.e., the percentage of task sets determined by each schedulability test condition to be schedulable. A higher acceptance ratio indicates a more accurate test, since all tests compared are safe but pessimistic. Figure 5.5 shows the acceptance ratio of the proposed test conditions ("TEST-1" is the general schedulability test presented in Sect. 5.4; "TEST-2" is the schedulability test for NP-FP in Sect. 5.5) and the simulation ("Sim").

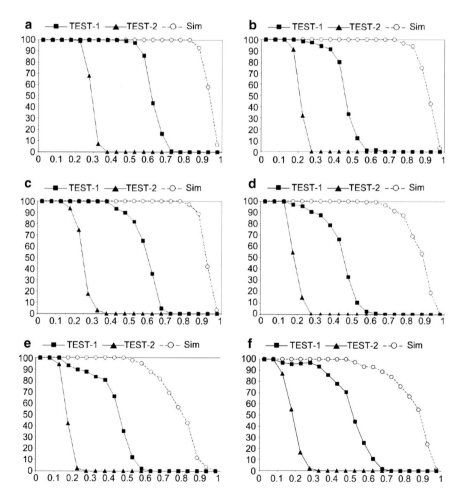

Fig. 5.5 Performance of the proposed schedulability tests in terms of acceptance ratio. (**a**) $U_i \in [0.1, 0.4]$, $T_i \in [10, 20]$; (**b**) $U_i \in [0.1, 0.6]$, $T_i \in [10, 20]$; (**c**) $U_i \in [0.1, 0.4]$, $T_i \in [10, 40]$; (**d**) $U_i \in [0.1, 0.6]$, $T_i \in [10, 40]$; (**e**) $U_i \in [0.1, 0.4]$, $T_i \in [10, 60]$ (**f**) $U_i \in [0.1, 0.6]$, $T_i \in [10, 60]$

In the simulation of each task set, we set all tasks release offsets as 0, task inter-release separation as T_i and task execution time as the C_i, and check the task set until its hyper-period. Even though these assumptions do not guarantee to generate the worst-case scenario in terms of schedulability, they are adopted since it is not computationally feasible to try all possible task release offsets, task inter-release times or task execution times. Simulation results obtained under this assumption may sometimes determine a task set to be schedulable even though it is not, but they can serve as an upper bound of the ratio of all schedulable task sets.

We follow the method in [135] to generate synthetic task sets: A task set of $M+1$ tasks is generated and tested for schedulability using the schedulability tests and simulation. Then the number of tasks is increased by 1 to generate a new task set, and it is tested again. This process is repeated until the total processor utilization exceeds M. The whole procedure is then repeated, starting with a new task set of $M+1$ tasks, until 10,000 task sets have been generated and tested. This method of generating synthetic task sets produces a fairly uniform distribution of total utilization values (except very low utilization values, which are not of our interest anyway).

The task parameter settings in Fig. 5.5a are as follows: The number of processors is 4; for each task, its period T_i is uniformly distributed in $[10, 20]$; the ratio between its deadline D_i and period T_i is uniformly distributed in $[0.9, 1]$; its utilization $U_i = \frac{C_i}{T_i}$ is uniformly distributed in $[0.1, 0.4]$. In Fig. 5.5b–f, the parameter concerning U_i and T_i is tuned while all other parameters are kept unchanged. The general schedulability test [TEST-1]'s acceptance ratio is low, because very coarse approximation on the workload in the problem window is used in the test. However, as can be seen in Fig. 5.5, by exploring the scheduling algorithm specific properties and refining the definition of the problem window, the performance of the second test [TEST-2] is significantly improved.

5.8 Simulations

In the section, we compare the performance of the scheduling algorithms NP-FP and P-FP by simulations. For hard real-time systems, the performance of a test condition is usually considered to make more sense than the absolute performance of a scheduling algorithm itself. However, studying the performance characteristics of the scheduling algorithms will disclose their "potentials" and may inspire the development of new scheduling and analysis techniques.

We use the same task generation strategy as in the last section, by which one can get a fairly uniform distribution of total utilization values (except very low utilization values, which are not of our interest anyway). We use the well-known Deadline-Monotonic (DM) policy to assign task priorities.

Since exhaustive simulations of global multiprocessor scheduling with sporadic task systems are not computationally feasible [85, 86], we adopt non-exhaustive simulations in our experiments: For each task set, the initial release time of each task τ_i is randomly chosen from $[0, T_i]$; when a task τ_i has experienced its minimal inter-

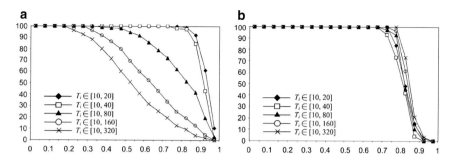

Fig. 5.6 Simulation experiments with different period ranges ($M = 4$, $U_i \in [0.1, 0.4]$). (**a**) $NP - FP$; (**b**) P-FP

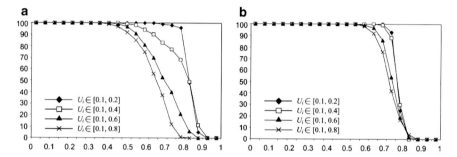

Fig. 5.7 Simulation experiments with different utilization ranges ($M = 8$, $T_i \in [10, 80]$). (**a**) $NP - FP$; (**b**) P-FP

release separation T_i from its last release time, at each time instant, the probability for τ_i to release the next job is $3/4$ (i.e., the probability is $1/4$ for τ_i to not release at this time instant, but wait until at least the next time instant). For each task set, the simulation starts from time 0, and terminates at $\min(5 \times 10^6, \text{hyper-period})$ (it is not feasible to simulate every task set to its hyper-period, since the task sets' hyper-periods could be very large under some parameter setting). For each parameter setting we simulate 5×10^5 task sets.

In the experiments of Fig. 5.6, the number of processor is 4, each task's utilization is randomly chosen from $[0.1, 0.4]$, and we tune *the range of task periods*. From Fig. 5.6a we can see that, as the range of task periods become broader, the performance of NP-FP degrades. This is because with a broader range of task periods (keeping the utilization range unchanged), tasks have larger execution times, and thereby cause longer non-preemptive blocking, which is bad for the schedulability of NP-FP. However, from Fig. 5.6b we can see that the period range has much smaller effect on the performance of P-FP.

In the experiments of Fig. 5.7, the number of processor is 8, each task's period is randomly chosen from $[10, 80]$, and we tune *the range of task utilizations*. Figure 5.7a shows that as tasks become "heavier," the performance of NP-FP

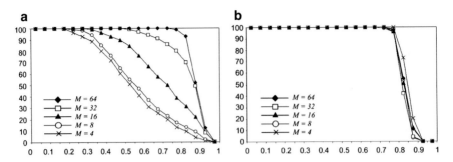

Fig. 5.8 Simulation experiments with different processor numbers ($T_i \in [10, 320]$, $U_i \in [0.1, 0.4]$). (**a**) $NP - FP$; (**b**) P-FP

degrades. This is because of the same reason as above: long execution times form longer non-preemptive blocking. Again, the effect of utilization range on P-FP is much smaller as shown in Fig. 5.7b.

In the experiments of Fig. 5.8, we tune *the number of processors* with a fixed period and utilization range ($T_i \in [10, 320]$, $U_i \in [0.1, 0.4]$). Figure 5.8a shows that NP-FP's performance gets better as the number of processors increases. Although a broad task period range causes longer non-preemptive blocking, the performance of NP-FP can be compensated by a larger processor number. This observation is of particular interest: Since the number of cores on a die is growing rapidly, the effect of non-preemptive blocking will become smaller and smaller in future "many-core" systems. Figure 5.8b shows that the effect of processor numbers on the performance of P-FP is very small.

Another interesting (even surprising) phenomenon disclosed by the above experiments is, under many parameter settings NP-FP actually performs better than P-FP (e.g., the experiments with $T_i \in [10, 20]$ and $T_i \in [10, 40]$ in Fig. 5.6, $U_i \in [0.1, 0.2]$ in Fig. 5.7, and $M = 64$ in Fig. 5.8), which counters the widely accepted intuition that non-preemptive scheduling in general has worse responsiveness than preemptive scheduling due to the non-preemptive blocking. Note that the context switch overhead is not considered in the simulation experiments. In the following we intuitively explain this counterintuitive phenomenon. In non-preemptive scheduling, a job will execute to completion once it starts, which is in general good for the schedulability of low-priority tasks. But at the same time, the low-priority tasks may form non-preemptive blocking, which is harmful for the schedulability of high-priority tasks. On single processor platforms, the effect of non-preemptive blocking is dominating: as long as there is a low-priority task with very long execution time, the high-priority tasks with "short" relative deadlines are doomed to miss their deadlines. However, this is not necessarily the case on multiprocessors: In global scheduling, even if there is a very "long" task non-preemptively occupying a processor, other "short" tasks still have chances to execute on other processors and meet their deadlines. As the number of processors increases, the effect of non-preemptive blocking becomes even smaller, as we can see in Fig. 5.8a. So when the

Table 5.2 An example task set is schedulable by NP-FP but not by P-FP

Task	T_i	D_i	C_i	Priority
τ_1	5	5	1	High
τ_2	5	5	1	Medium
τ_3	11	11	$8 + \epsilon$	Low

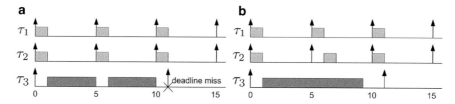

Fig. 5.9 Schedule of the task set in Table 5.2 under (**a**) P-FP and (**b**) NP-FP

effect of non-preemptive blocking is small, and the benefit of non-preemption for low-priority tasks is dominating, NP-FP will exhibit a better real-time performance than P-FP. As an example, we consider the task set in Table 5.2 running on 2 processors. Figure 5.9 shows the schedule under P-FP and NP-FP, respectively (assuming each task's initial release time is 0, and is released periodically). We can see in Fig. 5.9a that the lowest-priority task τ_3 misses its deadline in P-FP, while a lot of processor utilization is wasted (one processor is idle during [1, 5] and [6, 10]). However, as shown in Fig. 5.9b, τ_3 can meet its deadline in NP-FP (it executes to completion once it starts at time 2). At the same time, the non-preemptive blocking of τ_3 is not enough to prevent τ_1 and τ_2 from meeting their deadlines. Actually this task set can pass the test condition [TEST-2], so it is indeed schedulable by NP-FP (without the assumption on its release pattern in Fig. 5.9).

In summary, we can conclude that the real-time performance of NP-FP heavily depends on the parameter characteristics, while P-FP's performance is quite stable under different parameter settings. In general NP-FP has better real-time performance than P-FP when (1) the range of task periods/deadlines is narrow, (2) the task utilization is low, and (3) the number of processors is large. Since the number of cores on multi-core platforms is rapidly increasing, we suggest that non-preemptive scheduling may have better potential than preemptive scheduling for a considerable part of applications in future multi-core/many-core systems.

5.9 Conclusions

In this chapter, we study the schedulability problem of global non-preemptive scheduling on multiprocessor platforms. We conducted our simulation experiments empirically comparing the real-time performance of preemptive and non-preemptive global fixed-priority scheduling, by which we obtained interesting (even surprising) results suggesting that for a considerably part of applications on multiprocessor platforms, non-preemptive scheduling is actually a better choice than preemptive scheduling regarding the real-time performance. As technical contributions, we studied the schedulability analysis problem of non-preemptive fixed-priority scheduling (NP-FP) on multiprocessors. We firstly derived a linear-time general schedulability test condition that works on not only NP-FP, but also any other work-conserving non-preemptive scheduling algorithm. Then we improve the analysis and present a test condition of quadratic time-complexity for NP-FP, which has significant performance improvement compared to the first test condition.

Chapter 6
Liu and Layland's Utilization Bound

Liu and Layland discovered the famous utilization bound $N(2^{\frac{1}{N}} - 1)$ for fixed-priority scheduling on single-processor systems in the 1970s. Since then, it has been a long-standing open problem to find fixed-priority scheduling algorithms with the same bound for multiprocessor systems. In this chapter, we present a partitioning-based fixed-priority multiprocessor scheduling algorithm with Liu and Layland's utilization bound.

6.1 Introduction

Utilization bound is a well-known concept first introduced by Liu and Layland in their seminal paper [32]. Utilization bound can be used as a simple and practical way to test the schedulability of real-time task sets, as well as a good metric to evaluate the "quality" of a scheduling algorithm. It was shown that the utilization bound of *Rate Monotonic Scheduling* (RMS) on single processors is $N(2^{\frac{1}{N}} - 1)$. For simplicity of presentation we let $\Theta(N) = N(2^{\frac{1}{N}} - 1)$.

Multiprocessor scheduling is usually categorized into two paradigms [48]: *global scheduling*, in which each task can execute on any available processor in the run-time, and *partitioned scheduling* in which each task is assigned to a processor beforehand, and during the run-time each task can only execute on this particular processor. Although global scheduling on average utilizes computing resource better, the best known utilization bound of global fixed-priority scheduling is only 38 % [51], which is much lower than the best known result of partitioned fixed-priority scheduling 50 % [59]. Fifty percent is also known as the maximum utilization bound for both global and partitioned fixed-priority scheduling [50, 63]. Although there exist scheduling algorithms, like the pfair family [148, 149], offering utilization bounds of 100 %, these scheduling algorithms are not priority-based and incur much higher context-switch overhead [150].

© Springer International Publishing Switzerland 2016
N. Guan, *Techniques for Building Timing-Predictable Embedded Systems*,
DOI 10.1007/978-3-319-27198-9_6

Recently a number of works have been done on the *semi-partitioned scheduling*, which can exceed the maximum utilization bound 50 % of the partitioned scheduling. In semi-partitioned scheduling, most tasks are statically assigned to one fixed processor as in partitioned scheduling, while a few number of tasks are split into several subtasks, which are assigned to different processors. A recent work [151] has shown that the worst-case utilization bound of semi-partitioned fixed-priority scheduling can achieve 65 %, which is still lower than 69.3 % (the worst-case value of $\Theta(N)$ when N is increasing to the infinity). This gap is even larger with a smaller N.

In this chapter, we propose a new fixed-priority scheduling algorithm for multiprocessor systems based on semi-partitioned scheduling, whose utilization bound is $\Theta(N)$. The algorithm uses RMS on each processor, and has the same task splitting overhead as in previous work.

We first propose a semi-partitioned fixed-priority scheduling algorithm, whose utilization bound is $\Theta(N)$ for a class of task sets in which the utilization of each task is no larger than $\Theta(N)/(1 + \Theta(N))$. This algorithm assigns tasks in decreasing period order, and always selects the processor with the least workload assigned so far among all processors, to assign the next task. Then we remove the constraint on the utilization of each task, by introducing an extra task pre-assigning mechanism; the algorithm can achieve the utilization bound of $\Theta(N)$ for any task set.

6.2 Prior Work

Semi-partitioned scheduling has been studied with both EDF scheduling [66, 67, 69, 152–155] and fixed-priority scheduling [151, 156, 157].

The first semi-partitioned scheduling algorithm is EDF-fm [66] for soft real-time systems based on EDF scheduling. Andersson et al. proposed EKG [152] for hard real-time systems, in which split tasks are forced to execute in certain time slots. Later EKG was extended to sporadic and arbitrary deadline task systems [67, 153] with the similar idea. Kato et al. proposed EDDHP and EDDP [69, 154] in which split tasks are scheduled based on priority rather than time slots. The worst-case utilization bound of EDDP is 65 %. Later Kato et al. proposed EDF-WM, which can significantly reduce the context switch overhead against previous work.

There are relatively fewer works on the fixed-priority scheduling side. Kato et al. proposed RMDP [156] and DMPM [157], both with the worst-case utilization bound of 50 %, which is the same as the partitioned scheduling without task splitting. Recently, Lakshmanan et al. [151] proposed the algorithm PDMS_HPTS_DS, which can achieve the worst-case utilization bound of 65 %, and can achieve the bound 69.3 % for a special type of task sets that consist of "light" tasks. They also conducted case studies on an Intel Core 2 Duo processor to characterize the practical overhead of task-splitting, and showed that the cache overheads due to task-splitting can be expected to be negligible on multi-core platforms.

6.3 Basic Concepts

We first introduce the processor platform and task model. The multiprocessor platform consists of M identical processors $\{P_1, P_2, \ldots P_M\}$. A task set $\tau = \{\tau_1, \tau_2, \ldots, \tau_N\}$ consists of N independent tasks. Each task τ_i is a 2-tuple $\langle C_i, T_i \rangle$, where C_i is the worst-case execution time, T_i is the minimum inter-release separation (also called period). T_i is also τ_i's relative deadline.

Tasks in τ are sorted in non-decreasing period order, i.e., $j > i \Rightarrow T_j \geq T_i$. Since our proposed algorithms use rate-monotonic scheduling (RMS) as the scheduling algorithm on each processor, we can use the task indices to represent the task priorities, i.e., τ_i has higher priority than τ_j if and only if $i < j$. The *utilization* of each task τ_i is defined as $U_i = C_i/T_i$.

6.4 The First Algorithm SPA1

A significant difference between SPA1 and the algorithms in previous work is that SPA1 employs a "worst-fit" partitioning, while all previous algorithms employ a "first-fit" partitioning [151, 156, 157].

The basic procedure of "first-fit" partitioning is as follows: select a processor, and assign tasks to this processor as much as possible to fill its capacity, then pick the next processor and repeat the procedure. In contrast, the "worst-fit" partitioning always selects the processor with the minimal total utilization of tasks that have been assigned to it, so the occupied capacities of all processors are increased roughly "in turn."

The reason for us to prefer worst-fit partitioning is intuitively explained as follows. A subtask τ_i^k's actual deadline ($\triangle_i^k$) is shorter than τ_i's original deadline T_i, and the sum of the synthetic utilizations of all τ_i's subtasks is larger than τ_i's original utilization U_i, which is the key difficulty for semi-partitioned scheduling to achieve the same utilization bound as on single-processors. With worst-fit partitioning, the occupied capacity of all processors is increased "in turn," and task splitting only occurs when the capacity of a processor is completely "filled." Then, if one partitions all tasks in increasing priority order, the split tasks in worst-fit partitioning will generally have relatively high priority levels on each processor. This is good for the schedulability of the task set, since the tasks with high priorities usually have better chance to be schedulable, so they can tolerate the shortened deadlines better. Consider an extreme scenario: if one can make sure that all split tasks' subtasks have the highest priority on their host processors, then there is no need to consider the shortened deadlines of these subtasks, since, being of the highest priority level on each processor, they are schedulable anyway. Thus, as long as the split tasks with shorten deadlines do not cause any problem, Liu and Layland's utilization bound can be easily achieved. The philosophy behind our proposed algorithms is *making the split subtasks get as high priority as possible on each processor*.

In contrast, with the first-fit partitioning, a split subtask may get quite low priority on its host processors.[1] For instance, with the algorithm in [151] that achieves the utilization bound of 65 %, in the worst case the second subtask of a split task will always get the lowest priority on its host processor.

As will be seen later in this section, SPA1 does not completely solve the problem. More precisely, SPA1 is restricted to a class of *light* task sets, in which the utilization of each task is no larger than $\Theta(N)/(1 + \Theta(N))$. Intuitively, this is because if a task's utilization is very large, its tail subtask might still get a relatively low priority on its host processor, even using worst-fit partitioning. (We will solve this problem with SPA2 in Sect. 6.5.)

In the following, we will introduce SPA1 as well as its utilization bound property. The remaining part of this section is structured as follows: we first present the partitioning algorithm of SPA1, and show that any task set τ satisfying $U(\tau) \leq \Theta(N)$ can be successfully partitioned by SPA1. Then we introduce how the tasks assigned to each processor are scheduled. Next, we prove that if a *light* task set is successfully partitioned by SPA1, then all tasks can meet their deadlines under the scheduling algorithm of SPA1. Together, this implies that any *light* task set with $U(\tau) \leq \Theta(N)$ is schedulable by SPA1, and finally indicates the utilization bound of SPA1 is $\Theta(N)$ for *light* task sets.

```
1:  if U(τ) > Θ(N) then abort
2:  UQ := [τ_N^1, τ_{N-1}^1, ..., τ_1^1]
3:  Ψ[1...M] := all zeros
4:  while UQ ≠ ∅ do
5:      P_q := the processor with the minimal Ψ
6:      τ_i^k := pop_front(UQ)
7:      if (U_i^k + Ψ[q] ≤ Θ(N)) then
8:          τ_i^k ↦ P_q
9:          Ψ[q] := Ψ[q] + U_i^k
10:     else
11:         split τ_i^k into two parts τ_i^k and τ_i^{k+1} such that
                 U_i^k + Ψ[q] = Θ(N)
12:         τ_i^k ↦ P_q
13:         Ψ[q] := Θ(N)
14:         push_front(τ_i^{k+1}, UQ)
15:     end if
16: end while
```

Algorithm 1: The partitioning algorithm of SPA1

---

[1] Under the algorithms in [157], a split subtask's priority is artificially advanced to the highest level on its host processor, which breaks down the RMS priority order and thereby leads to a lower utilization bound.

6.4.1 **SPA1**: *Partitioning and Scheduling*

The partitioning algorithm of SPA1 is very simple, which can be briefly described as follows:

- We assign tasks in increasing priority order, and always select the processor on which the total utilization of tasks have been assigned so far is minimal among all the processors.
- When a task (subtask) cannot be assigned entirely to the current selected processor, we split it into two parts and assign the first part such that the total utilization of the current selected processor is $\Theta(N)$, and assign the second part to the next selected processor.

The precise description of the partitioning algorithm is in Algorithm 6.4. UQ is the list accommodating unassigned tasks, sorted in increasing priority order. UQ is initialized by $\{\tau_N^1, \tau_{N-1}^1, \ldots, \tau_1^1\}$, in which each element $\tau_i^1 = \langle c_i^1 = C_i, T_i, \Delta_k^1 = T_i \rangle$ is the initial subtask form of task τ_i. Each element $\Psi[q]$ in the array $\Psi[1 \ldots M]$ denotes the sum of the utilization of tasks that have been assigned to processor P_q.

The work flow of SPA1 is as follows. In each loop iteration, we pick the task at the front of UQ, denoted by τ_i^k, which has the lowest priority among all unassigned tasks. We try to assign τ_i^k to the processor P_q, which has the minimal $\Psi[q]$ among all elements in $\Psi[1 \ldots M]$. If

$$U_i^k + \Psi[q] \leq \Theta(N)$$

then we can assign the entire τ_i^k to P_q, since there is enough capacity available on P_q. Otherwise, we split τ_i^k into two subtasks τ_i^k and τ_i^{k+1}, such that

$$U_i^k + \Psi[q] = \Theta(N)$$

(Note that with $U_i^k = c_i^k / T_i$ we denote the utilization of subtask τ_i^k.) We further set $\Psi[q] := \Theta(N)$, which means this processor P_q is *full* and we will not assign any more tasks to P_q.

Then we insert τ_i^{k+1} back to the front of UQ, to assign it in the next loop iteration. We continue this procedure until all tasks have been assigned.

It is easy to see that all task sets below the desired utilization bound can be successfully partitioned by SPA1:

Lemma 6.1. *Any task set with*

$$U(\tau) \leq \Theta(N) \tag{6.1}$$

can be successfully partitioned to M processors with SPA1.

Note that there is no schedulability guarantee in the partitioning algorithm. It will be proved in the next subsection.

After the tasks are assigned (and possibly split) to the processors by the partitioning algorithm of **SPA1**, they will be scheduled using RMS on each processor locally, i.e., with their original priorities. The subtasks of a split task respect their precedence relations, i.e., a split subtask τ_i^k is ready for execution when its preceding subtask τ_i^{k-1} on some other processor has finished.

6.4.2   Schedulability

We first show an important property of **SPA1**:

Lemma 6.2. *After partitioning according to* **SPA1**, *each body subtask has the highest priority on its host processor.*

Proof. In the partitioning algorithm of **SPA1**, task splitting only occurs when a processor is full. Thus, after a body task was assigned to a processor, there will be no more tasks assigned to it. Further, the tasks are partitioned in increasing priority order, so all tasks assigned to the processor before have lower priority.

By Lemma 6.2, we further know that the response time of each body subtask equals its execution time, so the synthetic deadline $\triangle_i^t$ of each tail subtask τ_i^t is calculated as follows:

$$\triangle_i^t = T_i - \sum_{j\in[1,B]} c_i^{bj} = T_i - (C_i - c_i^t) \tag{6.2}$$

So we can view the scheduling in **SPA1** on each processor without considering the synchronization between the subtasks of a split task, and just *regard every split subtask τ_i^k as an independent task with period T_i and a shorter relative deadline $\triangle_i^k$ calculated by Eq. (6.2)*, as shown in Fig. 6.1.

In the following we prove the schedulability of non-split tasks, body subtasks, and tail subtasks, respectively.

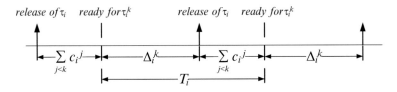

Fig. 6.1 Each subtask τ_i^k can be viewed as an independent task with period of T_i and deadline of $\triangle_i^k$

Non-split Tasks

Lemma 6.3. *If task set τ with $U(\tau) \leq \Theta(N)$ is partitioned by* SPA1, *then any* **non-split task** *of τ can meet its deadline.*

Proof. The tasks on each processor are scheduled by RMS, and the sum of the utilization of all tasks on a processor is no larger than $\Theta(N)$. Further, the deadlines of the non-split tasks are unchanged and therefore still equal their periods. Thus, each non-split task is schedulable.

Note that although the synthetic deadlines of other subtasks are shorter than their original periods, this does not affect the schedulability of the non-split tasks, since only the periods of these subtasks are relevant to the schedulability of the non-split tasks.

Body Subtasks

Lemma 6.4. *If task set τ with $U(\tau) \leq \Theta(N)$ is partitioned by* SPA1, *then any* **body subtask** *of τ can meet its deadline.*

Proof. The body subtasks have the highest priorities on their host processors and will therefore always meet their deadlines. (This holds even though the deadlines were shortened because of the task splitting).

Tail Subtasks

Now we prove the schedulability for an arbitrary tail subtask τ_i^t, during which we only focus on τ_i^t, but do not consider whether other tail subtasks are schedulable or not. Since the same reasoning can be applied to every tail subtask, the proofs guarantee that all tail subtasks are schedulable.

Suppose task τ_i is split into B body subtasks and one tail subtask. Recall that we use $\tau_i^{b_j}, j \in [1, B]$ to denote the jth body subtask of τ_i, and τ_i^t to denote τ_i's tail subtask. $U_i^{b_j} = c_i^{b_j}/T_i$ and $U_i^t = c_i^t/T_i$ denotes $\tau_i^{b_j}$'s and τ_i^t's original utilization, respectively.

Additionally, we use the following notations (cf. Fig. 6.2):

- For each body subtask $\tau_i^{b_j}$, let X^{b_j} denote the sum of the utilizations of all the tasks τ_k assigned to P^{b_j} with lower priority than $\tau_i^{b_j}$.
- For the tail subtask τ_i^t, let X^t denote the sum of the utilizations of all the tasks assigned to P^t with lower priority than τ_i^t.
- For the tail subtask τ_i^t, let Y^t denote the sum of the utilizations of all the tasks assigned to P^t with higher priority than τ_i^t.

We can use these now for the schedulability of the tail subtasks:

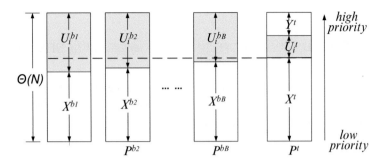

Fig. 6.2 Illustration of X^{b_j}, X^t, and Y^t

Lemma 6.5. *Suppose a tail subtask τ_i^t is assigned to processor P_t. If τ_i^t satisfies*

$$Y^t \cdot T_i / \triangle_i^t + V_i^t \le \Theta(N), \tag{6.3}$$

then τ_i^t can meet its deadline.

Proof. The proof idea is as follows: We consider the set Γ consisting of τ_i^t and all tasks with higher priority than τ_i^t *on the same processor*, i.e., the tasks contributing to Y^t. For this set, we construct a new task set $\tilde{\Gamma}$, in which the tasks' periods that are larger than $\triangle_i^t$ are all reduced to $\triangle_i^t$. The main idea is to first show that the counterpart of τ_i^t is schedulable with this new set $\tilde{\Gamma}$ by RMS because of the utilization bound, and then to prove this implies the schedulability of τ_i^t in the original set Γ.

In particular, let P_t be the processor to which τ_i^t is assigned. We define Γ as follows:

$$\Gamma = \{\tau_h^k \mid \tau_h^k \mapsto P_t \wedge h \le i\} \tag{6.4}$$

We now give the construction of $\tilde{\Gamma}$: For each task $\tau_h^k \in \Gamma$, we have a counterpart $\tilde{\tau}_h^k$ in $\tilde{\Gamma}$. The only difference is that we possibly reduce the periods:

$$\tilde{c}_h^k = c_h^k, \qquad \tilde{T}_h = \begin{cases} T_h, & \text{if } T_h \le \triangle_i^t \\ \triangle_i^t, & \text{if } T_h > \triangle_i^t \end{cases}$$

We also keep the same priority order of tasks in $\tilde{\Gamma}$ as their counterparts in Γ, which is still a rate-monotonic ordering.

Figure 6.3 illustrates the construction. In Fig. 6.3a, Γ contains three tasks. τ_1 has a period that is smaller than $\triangle_i^t$, and τ_2 has a larger one. Further, τ_i^t is contained in Γ. According to the construction, $\tilde{\Gamma}$ in Fig. 6.3b has also three tasks $\tilde{\tau}_1$, $\tilde{\tau}_2$, and $\tilde{\tau}_i^t$, where only the periods of $\tilde{\tau}_2$ and $\tilde{\tau}_i^t$ are reduced to $\triangle_i^t$.

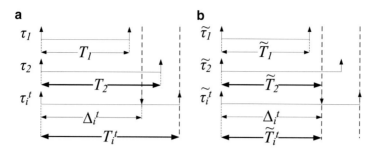

Fig. 6.3 Illustration of (**a**) Γ and (**b**) $\tilde{\Gamma}$

Now we show the schedulability of $\tilde{\tau}_i^t$ in $\tilde{\Gamma}$. We do this by showing the sufficient upper bound of $\Theta(N)$ on the total utilization of $\tilde{\Gamma}$.

$$U(\tilde{\Gamma}) = \sum_{\tau_h^k \in \Gamma} c_h^k / \widetilde{T}_h = \sum_{\tau_h^k \in \Gamma \setminus \{\tau_i^t\}} c_h^k / \widetilde{T}_h + V_i^k \qquad (6.5)$$

We now do a case distinction for tasks $\widetilde{\tau}_h^k \in \tilde{\Gamma}$, according to whether their periods were reduced or not.

- If $T_h \le \Delta_i^t$, we have $\widetilde{T}_h = T_h$. Since $T_i > \Delta_i^t$, we have

$$c_h^k / \widetilde{T}_h = c_h^k / T_h = U_h^k < U_h^k \cdot T_i / \Delta_i^t$$

- If $T_h > \Delta_i^t$, we have $\widetilde{T}_h = \Delta_i^t$. Because of the priority ordered by periods, we have $T_h \le T_i$. Thus:

$$c_h^k / \widetilde{T}_h = c_h^k / \Delta_i^t \le c_h^k / T_h \cdot T_i / \Delta_i^t = U_h^k \cdot T_i / \Delta_i^t$$

Both cases lead to $c_h^k / \widetilde{T}_h \le U_h^k \cdot T_i / \Delta_i^t$, so we can apply this to (6.5) from above:

$$U(\widetilde{\Gamma}) \le \sum_{\tau_h^k \in \Gamma \setminus \{\tau_i^t\}} U_h^k \cdot T_i / \Delta_i^t + V_i^k \qquad (6.6)$$

Since $Y^t = \sum_{\tau_h^k \in \Gamma \setminus \{\tau_i^t\}} U_h^k$, we have

$$U(\tilde{\Gamma}) \le Y^t \cdot T_i / \Delta_i^t + V_i^t$$

Finally, by the assumption from Condition (6.3) we know that the right-hand side is at most $\Theta(N)$, and thus $U(\tilde{\Gamma}) \le \Theta(N)$. Therefore, $\widetilde{\tau}_i^k$ is schedulable. Note that in $\tilde{\Gamma}$ there could exist other tail subtasks whose deadlines are shorter than their periods.

However, this does not invalidate that the condition $U(\tilde{\Gamma}) \leq \Theta(N)$ is sufficient to guarantee the schedulability of $\widetilde{\tau_i^t}$ under RMS.

Now we need to see that this implies the schedulability of τ_i^t. Recall that the only difference between Γ and $\tilde{\Gamma}$ is that the period of a task in Γ is possibly larger than its counterpart in $\tilde{\Gamma}$. So the interference τ_i^t suffered from the higher-priority tasks in Γ is no larger than the interference $\widetilde{\tau_i^t}$ suffered in $\tilde{\Gamma}$, and since the deadlines of $\widetilde{\tau_i^t}$ and τ_i^t are the same, we know the schedulability of $\widetilde{\tau_i^t}$ implies the schedulability of τ_i^t.

It remains to show that Condition (6.3) holds, which was the assumption for this lemma and thus a sufficient condition for tail subtasks to be schedulable. As in the introduction of this section, this condition does not hold in general for SPA1, but only for certain *light* task sets:

Definition 6.1. A task τ_i is a *light* task if

$$U_i \leq \frac{\Theta(N)}{1 + \Theta(N)}.$$

Otherwise, τ_i is a *heavy* task.

A task set τ is a *light* task sets if all tasks in τ are light tasks.

Lemma 6.6. *Suppose a tail subtask τ_i^t is assigned to processor P_t. If τ_i is a light task, we have*

$$Y^t \cdot T_i / \Delta_i^t + V_i^t \leq \Theta(N).$$

Proof. We will first derive a general upper bound on Y^t based on the properties of X^{b_j}, X^t and the subtasks' utilizations. Based on this, we derive the bound we want to show, using the assumption that τ_i is a light task.

For deriving the upper bound on Y^t, we note that as soon as a task is split into a body subtask and a rest, the processor hosting this new body subtask is full, i.e., its utilization is $\Theta(N)$. Further, each body subtask has by construction the highest priority on its host processor, so we have

$$\forall j \in [1, B] : U_i^{b_j} + X^{b_j} = \Theta(N)$$

We sum over all B of these equations, and get

$$\sum_{j \in [1,B]} U_i^{b_j} + \sum_{j \in [1,B]} X^{b_j} = B \cdot \Theta(N) \tag{6.7}$$

Now we consider the processor containing τ_i^t, denoted by P_t. Its total utilization is $X^t + U_i^t + Y^t$ and is at most $\Theta(N)$, i.e.,

$$X^t + U_i^t + Y^t \leq \Theta(N).$$

We combine this with (6.7) and get

$$Y^t \leq \frac{\sum_{j \in [1,B]} U_i^{b_j}}{B} + \frac{\sum_{j \in [1,B]} X^{b_j}}{B} - U_i^t - X^t \tag{6.8}$$

In order to simplify this, we recall that during the partitioning phase, we always select the processor with the smallest total utilization of tasks that have been assigned to it so far. (Recall line 5 in Algorithm 6.4). This implies $X^{b_j} \leq X^t$ for all subtasks $\tau_i^{b_j}$. Thus, the sum over all X^{b_j} is bounded by $B \cdot X^t$ and we can cancel out both terms in (6.8):

$$Y^t \leq \frac{\sum_{j \in [1,B]} U_i^{b_j}}{B} - U_i^t$$

Another simplification is possible using that $B \geq 1$ and that τ_i's utilization U_i is the sum of the utilizations of all of its subtasks, i.e., $\sum_{j \in [1,B]} U_i^{b_j} = U_i - U_i^t$:

$$Y^t \leq U_i - 2 \cdot U_i^t$$

We are now done with the first part, i.e., deriving an upper bound for Y^t. This can easily be transformed into an upper bound on the term we are interested in:

$$Y^t \cdot \frac{T_i}{\Delta_i^t} + V_i^t \leq (U_i - 2 \cdot U_i^t) \cdot \frac{T_i}{\Delta_i^t} + V_i^t \tag{6.9}$$

For the rest of the proof, we try to bound the right-hand side from above by $\Theta(N)$ which will complete the proof. The key is to bring it into a form that is suitable to use the assumption that τ_i is a *light* task.

As a first step, we use that the synthetic deadline of τ_i^t is the period T_i reduced by the total computation time of τ_i's body subtasks, i.e., $\Delta_i^t = T_i - (C_i - c_i^t)$, cf. Eq. (6.2). Further, we use the definitions $U_i = C_i/T_i$, $U_i^t = c_i^t/T_i$ and $V_i^t = c_i^t/\Delta_i^t$ to derive

$$(U_i - 2 \cdot U_i^t) \cdot \frac{T_i}{\Delta_i^t} + V_i^t = \frac{C_i - c_i^t}{T_i - (C_i - c_i^t)}$$

Since $c_i^t > 0$, we can find a simple upper bound of the right-hand side:

$$\frac{C_i - c_i^t}{T_i - (C_i - c_i^t)} = \frac{T_i}{T_i - (C_i - c_i^t)} - 1 < \frac{T_i}{T_i - C_i} - 1$$

Since τ_i is a *light* task, we have

$$U_i \leq \frac{\Theta(N)}{1 + \Theta(N)}$$

and by applying $U_i = C_i/T_i$ to the above, we can obtain

$$\frac{T_i}{T_i - C_i} - 1 \le \Theta(N)$$

Thus, we have established that $\Theta(N)$ is an upper bound of $Y^t \cdot \frac{T_i}{\Delta_i^t} + V_i^t$ with which we started in (6.9).

From Lemmas 6.5 and 6.6 it follows directly the desired property:

Lemma 6.7. *If task set τ with $U(\tau) \le \Theta(N)$ is partitioned by* SPA1*, then any* **tail subtask** *of a* light *task of τ can meet its deadline.*

6.4.3   Utilization Bound

By Lemma 6.1 we know that a task set τ can be successfully partitioned by the partitioning algorithm of SPA1 if $U(\tau)$ is no larger than $\Theta(N)$. If τ has been successfully partitioned, by Lemmas 6.3 and 6.4 we know that all the non-split task and body subtasks are schedulable. By Lemma 6.7 we know a tail subtask τ_i^k is also schedulable if τ_i is a light task. Since, in general, it is a priori unknown which tasks will be split, we pose this constraint of being light to all tasks in τ to have a sufficient schedulability test condition:

Theorem 6.1. *Let τ be a task set only containing* light *tasks. τ is schedulable with* SPA1 *on M processors if*

$$U(\tau) \le \Theta(N) \tag{6.10}$$

In other words, the utilization bound of SPA1 is $\Theta(N)$ for task sets only containing tasks with utilization no larger than $\Theta(N)/(1 + \Theta(N))$.

$\Theta(N)$ is a decreasing function with respect to N, which means the utilization bound is higher for task sets with fewer tasks. We use N^* to denote the maximal number of tasks (subtasks) assigned to on each processor, so $\Theta(N^*)$, which is strictly larger than $\Theta(N)$ also serves as the utilization bound on each processor. Therefore we can use $\Theta(N^*)$ to replace $\Theta(N)$ in the derivation above, and get that the utilization bound of SPA1 is $\Theta(N^*)$ for task sets only containing tasks with utilization no larger than $\Theta(N^*)/(1 + \Theta(N^*))$. It is easy to see that there is at least one task assigned to each processor, and two subtasks of a task cannot be assigned to the same processor. Therefore the number of tasks executing one each processor is at most $N - M + 1$, which can be used as an over-approximation of N^*.

6.5   The Second Algorithm SPA2

In this section we introduce our second semi-partitioned fixed-priority scheduling algorithm SPA2, which has the utilization bound of $\Theta(N)$ for task sets without any constraint.

As discussed in the beginning of Sect. 6.4, the key point for our algorithms to achieve high utilization bounds is to make each split task getting a priority as high as possible on its host processor. With SPA1, the tail subtask of a task with very large utilization could have a relatively low priority on its host processor, as the example in Fig. 6.4 illustrates. This is why the utilization bound of SPA1 is not applicable to task sets containing heavy tasks.

To solve this problem, we propose the second semi-partitioned algorithm SPA2 in this section. The main idea of SPA2 is to pre-assign each heavy task whose tail subtask might get a low priority, before partitioning other tasks, therefore these heavy tasks will not be split.

Note that if one simply pre-assigns all heavy tasks, it is still possible for some tail subtask to get a low priority level on its host processor. Consider the task set in Table 6.1 with two processors, and for simplicity we assume $\Theta(N) = 0.8$, and $\Theta(N)/(1 + \Theta(N)) = 4/9$. If we pre-assign the heavy task τ_1 to processor P_1, then assign τ_2 and τ_3 by the partitioning algorithm of SPA1, the task partitioning looks as follows:

1. $\tau_1 \mapsto P_1$,
2. $\tau_3 \mapsto P_2$,
3. τ_2 cannot be entirely assigned to P_2, so it is split into two subtasks $\tau_2^1 = \langle 3.75, 10, 10 \rangle$ and $\tau_2^2 = \langle 0.5, 10, 6.25 \rangle$, and $\tau_2^1 \mapsto P_2$,
4. $\tau_2^2 \mapsto P_1$.

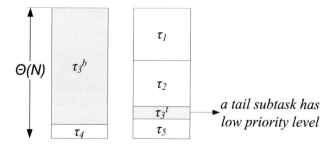

Fig. 6.4   The tail subtask of a task with large utilization may have a low priority level

Table 6.1   An example task set

Task	C_i	T_i	Heavy task?	Priority
τ_1	3	4	Yes	Highest
τ_2	4.25	10	No	Middle
τ_3	4.25	10	No	Lowest

Then the tasks on each processor are scheduled by RMS. We can see that the tail
subtask τ_2^2 has the lowest priority on P_1 and will miss its deadline due to the higher
priority task τ_1. However, if we do not pre-assign τ_1 and just do the partitioning with
SPA1, this task set is schedulable.

To overcome this problem, a more sophisticated pre-assigning mechanism is
employed in our second algorithm SPA2. Intuitively, SPA2 pre-assigns exactly
those heavy tasks for which pre-assigning them will not cause any tail subtask to
miss deadline. This is checked using a simple test. Those heavy tasks that don't
satisfy this test will be assigned (and possibly split) later together with the light
tasks. The key for this to work is, that for these heavy tasks, we can use the property
of failing the test in order to show that their tail subtasks will not miss the deadlines
either.

6.5.1   SPA2: Partitioning and Scheduling

We first introduce some notations. If a heavy task τ_i is pre-assigned to a processor
P_q in SPA2, we call τ_i as a *pre-assigned task*, otherwise a *normal task*, and call P_q
as a *pre-assigned processor*, otherwise a *normal processor*.

The partitioning algorithm of SPA2 contains three steps:

1. We first pre-assign the heavy tasks that satisfy a particular condition to one
   processor each.
2. We do task partitioning with the remaining (i.e., normal) tasks and remaining
   (i.e., normal) processors using SPA1 until all the normal processors are full.
3. The remaining tasks are assigned to the pre-assigned processors; the assignment
   selects one processor to assign as many tasks as possible, until it becomes full,
   then select the next processor.

The precise description of the partitioning algorithm of SPA2 is shown in
Algorithm 6.5. We first introduce the data structures used in the algorithm:

- PQ is the list of all processors. It is initially $[P_1, P_2, \ldots, P_M]$ and processors are
  always taken out and put back in the front.
- PQ_{pre} is the list to accommodate pre-assigned processors, initially empty.
- UQ is the list to accommodate the unassigned tasks after Step (1). Initially it is
  empty, and during Step (1), each task τ_i that is determined not to be pre-assigned
  will be put into UQ (already in its subtask form τ_i^1).
- $\Psi[1 \ldots M]$ is an array, which has the same meaning as in SPA1: each element
  $\Psi[q]$ in the array $\Psi[1 \ldots M]$ denotes the sum of the utilization of tasks that have
  been assigned to processor P_q.

In the following we use the task set example in Table 6.2 with four processors
to demonstrate how the partitioning algorithm of SPA2 works. For simplicity, we
assume $\Theta(N) = 0.7$, then the utilization threshold for light tasks $\Theta(N)/(1+\Theta(N))$
is around 0.41. The initial state of the data structures is as follows:

```
 1: if U(τ) > Θ(N) then abort
 2: PQ := [P₁, P₂, ..., P_M]
 3: PQ_pre := ∅
 4: UQ := ∅
 5: Ψ[1...M] := all zeros
 6: for i := 1 to N do
 7:     if τᵢ is heavy ∧
            ∑_{j>i} U_j ≤ (|PQ| − 1) · Θ(N) then
 8:         P_q := pop_front(PQ)
 9:         Pre-assign τᵢ to P_q
10:         push_front(P_q, PQ_pre)
11:         Ψ[q] := Ψ[q] + Uᵢ
12:     else
13:         push_front(τᵢ¹, UQ)
14:     end if
15: end for
16: while UQ ≠ ∅ do
17:     τᵢᵏ := pop_front(UQ)
18:     if ∃P_q ∈ PQ : Ψ[q] ≠ Θ(N) then
19:         P_q := the element in PQ with the minimal Ψ
20:     else
21:         P_q := pop_front(PQ_pre)
22:     end if
23:     if Uᵢᵏ + Ψ[q] ≤ Θ(N)  then
24:         τᵢᵏ ↦ P_q
25:         Ψ[q] := Ψ[q] + Uᵢᵏ
26:         if P_q came from PQ_pre then
27:             push_front(P_q, PQ_pre)
28:         end if
29:     else
30:         split τᵢᵏ into two parts τᵢᵏ and τᵢᵏ⁺¹ such that
            Uᵢᵏ + Ψ[q] = Θ(N)
31:         τᵢᵏ ↦ P_q
32:         Ψ[q] = Θ(N)
33:         push_front(τᵢᵏ⁺¹, UQ)
34:     end if
35: end while
```

Algorithm 2: The partitioning algorithm of SPA2

Table 6.2 An example
demonstrating SPA2

Task	C_i	T_i	Heavy task?	Priority
τ_1	0.5	10	No	highest
τ_2	4.5	10	Yes	
τ_3	6	10	Yes	
τ_4	4	10	No	
τ_5	3	10	No	
τ_6	6	10	Yes	
τ_7	3	10	No	Lowest

- $PQ = [P_1, P_2, P_3, P_4]$
- $PQ_{pre} = \emptyset$
- $UQ = \emptyset$
- $\Psi[1 \ldots 4] = [0, 0, 0, 0]$

In Step (1) (lines 6–15), each task τ_i in τ is visited in increasing index order, i.e., decreasing priority order. If τ_i is a heavy task, we evaluate the following condition (line 7):

$$\sum_{j>i} U_j \leq (|PQ| - 1) \cdot \Theta(N) \tag{6.11}$$

in which $|PQ|$ is the number of processors left in PQ so far. A heavy task τ_i is determined to be pre-assigned to a processor if this condition is satisfied. The intuition for this is: If we pre-assign this task τ_i, then there is enough space on the remaining processors to accommodate all remaining lower priority tasks. That way, no lower priority tail subtask will end up on the processor which we assign τ_i to.

In our example, we first visit the first task τ_1^1. It is a light task, so we put it to the front of UQ (line 13). The next task τ_2 is heavy, but Condition (6.11) with $|PQ| = 4$ is not satisfied, so we put τ_2^1 to the front of UQ. The next task τ_3 is heavy, and Condition (6.11) with $|PQ| = 4$ is satisfied. Thus, we pre-assign τ_3 to P_1, and put P_1 to the front of PQ_{pre} (lines 8–10). τ_4 and τ_5 are both light tasks, so we put them to UQ, respectively. τ_6 is heavy, and Condition (6.11) with $|PQ| = 3$ (P_1 has been taken out from PQ and put into PQ_{pre}) is satisfied, so we pre-assign τ_5 to P_2, and put P_2 to the front of PQ_{pre}. The last task τ_7 is light, so it is put to the front of UQ. So far, the Step (1) phase has been finished, and the state of the data structures is as follows:

- $PQ = [P_3, P_4]$
- $PQ_{pre} = [P_2, P_1]$
- $UQ = [\tau_7^1, \tau_5^1, \tau_4^1, \tau_2^1, \tau_1^1]$
- $\Psi[1 \ldots 4] = [0.6, 0.6, 0, 0]$

Note that the processors in PQ_{pre} are in decreasing priority order of the pre-assigned tasks on them, and the tasks in UQ are in decreasing priority order.

Steps (2) and (3) are both in the while loop of lines $16 \sim 35$. In Step (2), the remaining tasks (which are now in UQ) are assigned to normal processors (the ones in PQ). Only as soon as all processors in PQ are full, the algorithm enters Step (3), in which tasks are assigned to processors in PQ_{pre} (decision in lines 18–22).

The operation of assigning a task τ_i^k (lines 23–34) is basically the same as in **SPA1**. If τ_i^k can be entirely assigned to P_q without task splitting, then $\tau_i^k \mapsto P_q$ and $\Psi[q]$ is updated (lines 24–28). If P_q is a pre-assigned processor, P_q is put back to the front of PQ_{pre} (lines 26–28), so that it will be selected again in the next loop iteration, otherwise no putting back operation is needed since we never take out elements from PQ, but just select the proper one in it (line 19).

If τ_i^k cannot be assigned to P_q entirely, τ_i^k is split into a new τ_i^k and another subtask τ_i^{k+1}, such that P_q becomes full after the new τ_i^k being assigned to it, and then we put τ_i^{k+1} back to UQ (see lines 29–33).

Note that there is an important difference between assigning tasks to normal processors and to pre-assigned processors. When tasks are assigned to normal processors, the algorithm always selects the processor with the minimal Ψ (the same as in SPA1). In contrast, when tasks are assigned to pre-assigned processors, always the processor at the front of PQ_{pre} is selected, i.e., we assign as many tasks as possible to the processor in PQ_{pre} whose pre-assigned task has the lowest priority, until it is full. As will be seen later in the schedulability proof, this particular order of selecting pre-assigned processors, together with the evaluation of Condition (6.11), is the key to guarantee the schedulability of heavy tasks.

With our running example, the remaining tasks are first assigned to the normal processors P_3 and P_4 in the same way as by SPA1. Thus, $\tau_7^1 \mapsto P_3$, then $\tau_5^1 \mapsto P_4$, then $\tau_4^1 \mapsto P_3$, then τ_2^1 is split into $\tau_2^1 = \langle 4, 10, 10 \rangle$ and $\tau_2^2 = \langle 0.5, 10, 6 \rangle$, and $\tau_2^1 \mapsto P_4$. So far, all normal processors are full, and the state of the data structures is as follows:

- $PQ = [P_3, P_4]$ (both P_3 and P_4 are full)
- $PQ_{pre} = [P_2, P_1]$
- $UQ = [\tau_2^2, \tau_1^1]$
- $\Psi[1 \ldots 4] = [0.6, 0.6, 0.7, 0.7]$

Then the remaining tasks in UQ are assigned to the pre-assigned processors. At first $\tau_2^2 \mapsto P_2$, after which P_2 is not full and still at the front of PQ_{pre}. So the next task τ_1^1 is also assigned to P_2. There is no unassigned task any more, so the algorithm terminates.

It is easy to see that any task set below the desired utilization bound can be successfully partitioned by SPA2:

Lemma 6.8. *Any task set with*

$$U(\tau) \leq \Theta(N)$$

can be successfully partitioned to M processors with SPA2.

After describing the partitioning part of SPA2, we also need to describe the scheduling part. It is the same as SPA1: on each processor the tasks are scheduled by RMS, respecting the precedence relations between the subtasks of a split task, i.e., a subtask is ready for execution as soon as the execution of its preceding subtask has been finished. Note that under SPA2, each body subtask is also with the highest priority on its host processor, which is the same as in SPA1. So we can view the scheduling on each processor as the RMS with a set of independent tasks, in which each subtask's deadline is shortened by the sum of the execution time of all its preceding subtasks.

6.5.2   *Properties*

Now we introduce some useful properties of **SPA2**.

Lemma 6.9. *Let τ_i be a heavy task, and there are η pre-assigned tasks with higher priority than τ_i. Then we know*

- *If τ_i is a pre-assigned task, it satisfies*

$$\sum_{j>i} U_j \leq (M - \eta - 1) \cdot \Theta(N) \tag{6.12}$$

- *If τ_i is not a pre-assigned task, it satisfies*

$$\sum_{j>i} U_j > (M - \eta - 1) \cdot \Theta(N) \tag{6.13}$$

Proof. The proof directly follows the partitioning algorithm of **SPA2**.

Lemma 6.10. *Each pre-assigned task has the lowest priority on its host processor.*

Proof. Without loss of generality, we sort all processors in a list Q as follows: we first sort all pre-assigned processors in Q, in decreasing priority order of the pre-assigned tasks on them; then, the normal processors follow in Q in an arbitrary order. We use P_x to denote the xth processor in Q. Suppose τ_i is a heavy task pre-assigned to P_q.

τ_i is a pre-assigned task, and the number of pre-assigned task with higher priority than τ_i is $q - 1$, so by Lemma 6.9 we know the following condition is satisfied:

$$\sum_{j>i} U_j \leq (M - q) \cdot \Theta(N) \tag{6.14}$$

In the partitioning algorithm of **SPA2**, normal tasks are assigned to pre-assigned processors only when all normal processors are full, and the pre-assigned processors are selected in increasing priority order of the pre-assigned tasks on them, so we know only when the processors $P_{q+1} \ldots P_M$ are all full, normal tasks can be assigned to processor P_q. The total capacity of processors $P_{q+1} \ldots P_M$ are $(M - q) \cdot \Theta(N)$ (in our algorithms a processor is full as soon as the total utilization on it is $\Theta(N)$), and by (6.14), we know when we start to assign tasks to P_q, the tasks with lower priority than τ_i all have been assigned to processors $P_{q+1} \ldots P_M$, so all normal tasks (subtasks) assigned to P_q have higher priorities than τ_i.

Lemma 6.11. *Each body subtask has the highest priority on its host processor.*

Proof. Given a body subtask $\tau_i^{b_j}$ assigned to processor P_{b_j}. Since task splitting only occurs when a processor is full, and all the normal tasks are assigned in increasing priority order, we know $\tau_i^{b_j}$ has the highest priority among all normal tasks on P_{b_j}.

Additionally, by Lemma 6.10 we know that if P_{b_j} is a pre-assigned processor, the pre-assigned task on P_{b_j} also has lower priority than $\tau_i^{b_j}$. So we know P_{b_j} has the highest priority on P_{b_j}.

6.5.3 Schedulability

By Lemma 6.11 we know that under SPA2 each body subtask has the highest priority on its host processor, so we know all body subtasks are schedulable.

The scheduling algorithm of SPA2 is still RMS, and the deadline of a non-split task still equals to its period, so the schedulability of non-split tasks can be proved in the same way as in SPA1 (Lemma 6.3).

In the following we will prove the schedulability of tail subtasks. Suppose τ_i is split into B body subtasks and one tail subtask. Recall that we use $\tau_i^{b_j}, j \in [1, B]$ to denote the jth body subtask of τ_i, and τ_i^t to denote τ_i's tail subtask. X^t, Y^t, and X^{b_j} are defined the same as in Sect. 6.4.2.

First we recall Lemma 6.5, which is used to prove the schedulability of tail subtasks in SPA1: if a tail subtask τ_i^t satisfies

$$Y^t \cdot T_i / \Delta_i^t + V_i^t \leq \Theta(N) \qquad (6.15)$$

τ_i^t can meet its deadline. This conclusion also holds for SPA2, since the scheduling algorithm on SPA2 is also RMS, which is only the relevant property required by the proof of Lemma 6.5. So proving the schedulability of tail subtasks is reduced to proving Condition (6.15) for tail subtasks under SPA2.

We call τ_i^t a *tail-of-heavy* if τ_i is heavy, otherwise a *tail-of-light*. In the following we prove Condition (6.15) for τ_i^t in three cases:

1. τ_i^t is a tail-of-light, and P_t is a normal processor,
2. τ_i^t is a tail-of-light, and P_t is a pre-assigned processor,
3. τ_i^t is a tail-of-heavy.

Case (1) can be proved in the same way as in SPA1, since both the partitioning and scheduling algorithm of SPA2 *on normal processors* are the same as SPA1. Actually one can regard the partitioning and scheduling of SPA2 on normal processors as the partitioning and scheduling with a subset of tasks (those are assigned to normal processors) on a subset of processors (normal processors) of SPA1. So the schedulability of τ_i^t in this case can be proved by exactly the same reasoning as for Lemma 6.6.

Now we prove Case (2), where τ_i^t is a tail-of-light, and P_t is a pre-assigned processor.

Lemma 6.12. *Suppose τ_i^t is a tail-of-light assigned to a pre-assigned processor P_t under* **SPA2**. *We have*

$$Y^t \cdot T_i / \triangle_i^t + V_i^t \le \Theta(N)$$

Proof. By Lemma 6.10 we know τ_i^t has higher priority than the pre-assigned task of P_t, so X^t is no smaller than the utilization of this pre-assigned task. And since a pre-assigned task must be heavy, we have

$$X^t > \frac{\Theta(N)}{1 + \Theta(N)} \tag{6.16}$$

On the other hand, since τ_i is light, we know

$$\frac{C_i}{T_i} \le \frac{\Theta(N)}{1 + \Theta(N)}$$

We use c_i^B to denote the total execution time of all τ_i's body tasks. Since $c_i^B < C_i$ and $\Theta(N) < 1$, we have

$$c_i^B < \frac{1}{1 + \Theta(N)} \cdot T_i$$

$$\Leftrightarrow T_i(1 - \frac{1}{1 + \Theta(N)}) < T_i - c_i^B$$

$$\Leftrightarrow \frac{T_i}{T_i - c_i^B}(\Theta(N) - \frac{\Theta(N)}{1 + \Theta(N)}) < \Theta(N)$$

$$\Leftrightarrow \frac{T_i}{\triangle_i^t}(\Theta(N) - \frac{\Theta(N)}{1 + \Theta(N)} - U_i^t) + V_i^t < \Theta(N) \tag{6.17}$$

By (6.17) and (6.16) we have

$$\frac{T_i}{\triangle_i^t}(\Theta(N) - X^t - U_i^t) + V_i^t < \Theta(N)$$

and since the total utilization on each processor is bounded by $\Theta(N)$, i.e.,

$$Y^t \le \Theta(N) - X^t - U_i^t$$

finally we have $Y^t \cdot T_i / \triangle_i^t + V_i^t < \Theta(N)$.

Now we prove Case (3), where τ_i^t is a tail-of-heavy. Note that in this case P_t can be either a pre-assigned or a normal processor.

Lemma 6.13. τ_i^t *is the tail subtask of a normal heavy task* τ_i, *then we have*

$$Y^t \cdot T_i/\triangle_i^t + V_i^t \le \Theta(N)$$

Proof. By the property in Lemma 6.9 concerning normal heavy tasks we know τ_i satisfies the condition

$$\sum_{j>i} U_j > (M - \eta - 1) \cdot \Theta(N)$$

in which η is the number of pre-assigned tasks with higher priority than τ_i.

We use $\mathscr{M}$ to denote the set of all processors, so $|\mathscr{M}| = M$, and use $\mathscr{H}$ to denote the set of the pre-assigned processors on which the pre-assigned tasks' priorities are higher than τ_i, so $|\mathscr{H}| = \eta$, so we have

$$\sum_{j>i} U_j > (M - |\mathscr{H}| - 1) \cdot \Theta(N) \tag{6.18}$$

By Lemma 6.10 we know any normal task assigned to a pre-assigned processor has higher priority than the pre-assigned task of this processor. Therefore, τ_i's body and tail subtasks are all assigned to processors in $\mathscr{M} \setminus \mathscr{H}$. Moreover, when we start to assign τ_i, all tasks with lower priority than τ_i have already been assigned (pre-assigned) to processors in $\mathscr{M} \setminus \mathscr{H}$, since pre-assigned tasks have already been assigned before dealing with the normal tasks, and all normal tasks are assigned in increasing priority order.

We use $\mathscr{K}$ to denote the set of processors in $\mathscr{M} \setminus \mathscr{H}$ that contain neither τ_i's body nor tail subtask, and for each processor $P_k \in \mathscr{K}$ we use X^k to denote the total utilization of the tasks with lower priority than τ_i assigned to P_k. Then we have

$$X^t + \sum_{j\in[1,B]} X^{bj} + \sum_{k\in[1,|\mathscr{K}|]} X^k = \sum_{j>i} U_j$$

Since $|\mathscr{K}| = M - |\mathscr{H}| - (B+1)$, and $\forall P_k \in \mathscr{K}, X_k \le \Theta(N)$, we have

$$X^t + \sum_{j\in[1,B]} X^{bj} \ge \sum_{j>i} U_j - (M - |\mathscr{H}| - (B+1)) \cdot \Theta(N) \tag{6.19}$$

By Inequalities (6.18) and (6.19) we have

$$X^t + \sum_{j\in[1,B]} X^{bj} > B \cdot \Theta(N) \tag{6.20}$$

Now we look at processor P_t, the total utilization on which is bounded by $\Theta(N)$, so we have

$$Y^t \le \Theta(N) - X^t - U_i^t \tag{6.21}$$

By (6.20) and (6.21) we have

$$Y^t \leq \Theta(N) - (B \cdot \Theta(N) - \sum_{j \in [1,B]} X^{b_j}) - U_i^t$$

and since $U_i^t + \sum_{j \in [1,B]} U_i^{b_j} = U_i$, we have

$$Y^t \leq \Theta(N) - B \cdot \Theta(N) - U_i + \left(\sum_{j \in [1,B]} X^{b_j} + \sum_{j \in [1,B]} U_i^{b_j} \right) \tag{6.22}$$

Since each body task has the highest priority on its host processor, and the total utilization of any processor containing a body subtask is $\Theta(N)$, we have

$$\sum_{l \in [1,B]} X^{b_l} + \sum_{l \in [1,B]} U_i^{b_l} = B \cdot \Theta(N) \tag{6.23}$$

By (6.22) and (6.23) we have

$$Y^t \leq \Theta(N) - U_i$$
$$\Leftrightarrow Y^t \cdot T_i/\triangle_i^t + V_i^t \leq (\Theta(N) - U_i) \cdot T_i/\triangle_i^t + V_i^t$$

By applying $U_i = C_i/T_i$ and $V_i^t = c_i^t/\triangle_i^t$ to the RHS of the above inequality, we get

$$Y^t \cdot T_i/\triangle_i^t + V_i^t \leq \Theta(N) \cdot T_i/\triangle_i^t - C_i/\triangle_i^t + c_i^t/\triangle_i^t \tag{6.24}$$

We use c_i^B to denote the sum of the execution time of all τ_i's body subtasks, so we have $c_i^t + c_i^B = C_i$ and $\triangle_i^t = T_i - c_i^B$. We apply these to the RHS of (6.24) and get

$$Y^t \cdot T_i/\triangle_i^t + V_i^t \leq \frac{T_i\Theta(N) - c_i^B}{T_i - c_i^B} \tag{6.25}$$

Since $\Theta(N) < 1$, we have

$$c_i^B > c_i^B \cdot \Theta(N)$$
$$\Leftrightarrow T_i \cdot \Theta(N) - c_i^B < T_i \cdot \Theta(N) - c_i^B \cdot \Theta(N)$$
$$\Leftrightarrow \frac{T_i \cdot \Theta(N) - c_i^B}{T_i - c_i^B} < \Theta(N) \tag{6.26}$$

So by Inequalities (6.25) and (6.26) we have

$$Y^t \cdot T_i/\triangle_i^t + V_i^t < \Theta(N)$$

6.5.4 Utilization Bound

Now we have known that any task set τ with $U(\tau) \leq \Theta(N)$ can be successfully partitioned on M processors by SPA2 (Lemma 6.8). In the last subsection, we have shown that under the scheduling algorithm of SPA2, the body subtasks are schedulable since they are always with the highest priority level on their host processors; the non-split tasks are also schedulable since the utilization on each processor is bounded by $\Theta(N)$. The schedulability for the tail subtasks is proved by case distinction, in which the schedulability for the *light* tail subtasks on *normal* processors can be proved by the same reasoning as for Lemma 6.6, for the *light* tail subtasks on *pre-assigned* processors is proved by Lemma 6.12, and for the *heavy* tail subtasks is proved by Lemma 6.13. So we have the following theorem:

Theorem 6.2. τ *is schedulable by* SPA2 *on M processors if*

$$U(\tau) \leq \Theta(N)$$

So $\Theta(N)$ is the utilization bound of SPA2 for any task set.

For the same reason as presented at the end of Sect. 6.4.2, we can use $\Theta(N^*)$, the maximal number of tasks (subtasks) assigned to on each processor, to replace $\Theta(N)$ in Theorem 6.2.

6.5.5 Task Splitting Overhead

With the algorithms proposed in this chapter, a task could be split into more than two subtasks. However, since the task splitting only occurs when a processor is full, for any task set that is schedulable by SPA2, the number of task splitting is at most $M - 1$, which is the same as in previous semi-partitioned fixed-priority scheduling algorithms [151, 156, 157], and as shown in case studies conducted in [151], this overhead can be expected to be negligible on multi-core platforms.

6.6 Conclusions and Future Work

In this chapter, we have developed a semi-partitioned fixed-priority scheduling algorithm for multiprocessor systems, with the well-known Liu and Layland's utilization bound for RMS on single processors. The algorithm enjoys the following property. If the utilization bound is used for the schedulability test, and a task set is determined schedulable by fixed-priority scheduling on a single processor of speed M, it is also schedulable by our algorithm on M processors of speed 1 (under the assumption that each task's execution time on the processors of speed 1 is still smaller than its deadline). Note that the utilization bound test is only sufficient but not necessary. As future work, we will challenge the problem of constructing algorithms holding the same property with respect to the exact schedulability analysis.

Chapter 7
Parametric Utilization Bounds

Future embedded real-time systems will be deployed on multi-core processors to meet the dramatically increasing high-performance and low-power requirements. This trend appeals to generalize established results on uniprocessor scheduling, particularly the various utilization bounds for schedulability test used in system design, to the multiprocessor setting. Recently, this has been achieved for the famous Liu and Layland utilization bound by applying novel task splitting techniques. However, *parametric* utilization bounds that can guarantee higher utilizations (up to 100 %) for common classes of systems are not yet known to be generalizable to multiprocessors as well. In this chapter, we solve this problem for most parametric utilization bounds by proposing new task partitioning algorithms based on exact response time analysis (RTA). In addition to the worst-case guarantees, as the exact RTA is used for task partitioning, our algorithms significantly improve average-case utilization over previous work.

7.1 Introduction

It has been widely accepted that future embedded real-time systems will be deployed on multi-core processors, to satisfy the dramatically increasing high-performance and low-power requirements. This trend demands effective and efficient techniques for the design and analysis of real-time systems on multi-cores.

A central problem in the real-time system design is *timing analysis*, which examines whether the system can meet all the specified timing requirements. Timing analysis usually consists of two steps: task-level timing analysis, which, for example, calculates the worst-case execution time of each task independently, and system-level timing analysis (also called *schedulability analysis*), which determines whether all the tasks can co-exist in the system and still meet all the time requirements.

© Springer International Publishing Switzerland 2016
N. Guan, *Techniques for Building Timing-Predictable Embedded Systems*,
DOI 10.1007/978-3-319-27198-9_7

One of the most commonly used schedulability analysis approaches is based on the *utilization bound*, which is a safe threshold of the system's workload: under this threshold the system is guaranteed to meet all the time requirements. The utilization-bound-based schedulability analysis is very efficient, and is especially suitable to embedded system design flow involving iterative design space exploration procedures. A well-known utilization bound is the $N(2^{1/N} - 1)$ bound for RMS (Rate Monotonic Scheduling) on uni-processors, discovered by Liu and Layland in the 1970s [32]. Recently, this bound has been generalized to multiprocessors scheduling by a partitioning-based algorithm [68].

The Liu and Layland utilization bound (*L&L* bound for short) is pessimistic: There are a significant number of task systems that exceed the *L&L* bound but are indeed schedulable. This means that system resources would be considerably under-utilized if one only relies on the *L&L* bound in system design.

If more information about the task system is available in the design phase, it is possible to derive higher *parametric* utilization bounds regarding known task parameters. A well-known example of parametric utilization bounds is the 100 % bound for *harmonic* task sets [158]: If the total utilization of a harmonic task set τ is no greater than 100 %, then every task in τ can meet its deadline under RMS on a uni-processor platform. Even if the whole task system is not harmonic, one can still obtain a significantly higher bound by exploring the "harmonic chains" of the task system [45]. In general, during the system design, it is usually possible to employ higher utilization bounds with available task parameter information to better utilize the resources and decrease the system cost. As will be introduced in Sect. 7.4, quite a number of higher parametric utilization bounds regarding different task parameter information have been derived for uni-processor scheduling.

This naturally raises an interesting question: Can we generalize these higher parametric utilization bounds derived for uni-processor scheduling to multiprocessors? For example, given a harmonic task system, can we guarantee the schedulability of the task system on a multiprocessor platform of M processors, if the utilization sum of all tasks is no larger than M?

In this chapter, we will address the above question by proposing new RMS-based partitioned scheduling algorithms (with task splitting). Generalizing the parametric utilization bounds from uni-processors to multiprocessors is challenging, even with the insights from our previous work generalizing the *L&L* bound to multiprocessor scheduling. The reason is that task splitting may "create" new tasks that do not comply with the parameter properties of the original task set, and thus invalidate the parametric utilization bound specific to the original task set's parameter properties. Section 7.4 presents this problem in detail. The main contribution of this chapter is a solution to this problem, which generalizes most of the parametric utilization bounds to multiprocessors.

The approach of this chapter is generic in the sense that it works irrespective of the form of the parametric utilization bound in consideration. The only restriction is a threshold on the parametric utilization bound value when some task has a large individual utilization; apart from that, any parametric utilization bound derived for single-processor RMS can be used to guarantee the schedulability of

multiprocessors systems via our algorithms. More specifically, we first proposed
an algorithm generalizing all known parametric utilization bounds for **RMS** to
multiprocessors, for a class of "light" task sets in which each task's individual
utilization is at most $\frac{\Theta(\tau)}{1+\Theta(\tau)}$, where $\Theta(\tau) = N(2^{1/N} - 1)$ is the *L&L* bound for
task set τ. Then we proposed the second algorithm that works for any task set and
all parametric utilization bounds under the threshold $\frac{2\Theta(\tau)}{1+\Theta(\tau)}$.[1]

Besides the improved utilization bounds, another advantage of our new algorithms is the significantly improved average-case performance. Although the algorithm in last chapter can achieve the *L&L* bound, it has the problem that it never
utilizes more than the worst-case bound. The new algorithms in this chapter use
exact analysis, i.e., RTA, instead of the utilization bound threshold as in the
algorithm of last chapter, to determine the maximal workload on each processor.
It is well known that on uni-processors, by exact schedulability analysis, the average
breakdown utilization of **RMS** is around 88 % [159], which is much higher than its
worst-case utilization bound 69.3 %. Similarly, our new algorithm has much better
performance than the algorithm in [68].

7.2 Related Work

Multiprocessor scheduling is usually categorized into two paradigms [48]: *global
scheduling*, where each task can execute on any available processor at run-time, and
partitioned scheduling, where each task is assigned to a processor beforehand, and
at run-time each task only executes on its assigned processor. Global scheduling
on average utilizes the resources better. However, the standard **RMS** and **EDF**
global scheduling strategies suffer from the Dhall effect [58], which may cause
a task system with arbitrarily low utilization to be unschedulable. Although the
Dhall effect can be mitigated by, e.g., assigning higher priorities to tasks with higher
utilizations as in **RM-US** [50], the best known utilization bound of global scheduling
is still quite low: 38 % for fixed-priority scheduling [51] and 50 % for **EDF**-
based scheduling [160]. On the other hand, partitioned scheduling suffers from the
resource waste similar to the bin-packing problem: the worst-case utilization bound
for any partitioned scheduling cannot exceed 50 %. Although there exist scheduling
algorithms like the Pfair family [148, 149], the LLREF family [161, 162] and the
EKG family [67, 152], offering utilization bounds up to 100 %, these algorithms
incur much higher context-switch overhead than priority-driven scheduling, which
is unacceptable in many real-life systems.

Recently, a number of works [66–69, 151, 152, 155–157] have studied partitioned
scheduling with *task splitting*, which can overcome the 50 % limit of the strict
partitioned scheduling. In this class of scheduling algorithms, while most tasks
are assigned to a fixed processor, some tasks may be (sequentially) divided into

---

[1]When N goes to infinity, $\Theta(\tau) \doteq 69.3\,\%$, $\frac{\Theta(\tau)}{1+\Theta(\tau)} \doteq 40.9\,\%$ and $\frac{2\Theta(\tau)}{1+\Theta(\tau)} \doteq 81.8\,\%$.

several parts and each part is assigned and thereby executed on a different (but fixed) processor. In this category, the utilization bound of the state-of-the-art EDF-based algorithm is 65 % [69], and our recent work [68] has achieved the *L&L* bound (in the worst case 69.3 %) for fixed-priority based algorithms.

7.3   Basic Concepts

We consider a multiprocessor platform consisting of M processors $\mathscr{P} = \{P_1, P_2, \ldots P_M\}$. A task set $\tau = \{\tau_1, \tau_2, \ldots, \tau_N\}$ complies with the *L&L* task model: Each task τ_i is a 2-tuple $\langle C_i, T_i \rangle$, where C_i is the worst-case execution time and T_i is the minimal inter-release separation (also called period). T_i is also τ_i's relative deadline. We use the RMS strategy to assign priorities: tasks with shorter periods have higher priorities. Without loss of generality we sort tasks in non-decreasing period order, and can therefore use the task indices to represent task priorities, i.e., $i < j$ implies that τ_i has higher priority than τ_j. The *utilization* of each task τ_i is defined as $U_i = C_i/T_i$, and the *total utilization* of task set τ is $\mathscr{U}(\tau) = \sum_{i=1}^{N} U_i$. We further define the *normalized utilization* of a task set τ on a multiprocessor platform with M processors:

$$\mathscr{U}_M(\tau) = \sum_{\tau_i \in \tau} U_i/M$$

Note that the subscript M in $\mathscr{U}_M(\tau)$ reminds us that the sum of all tasks' utilizations is divided by the number of processors M.

A partitioned scheduling algorithm (with task splitting) consists of two parts: the *partitioning algorithm*, which determines how to split and assign each task (or rather each of its parts) to a fixed processor, and the *scheduling algorithm*, which determines how to schedule the tasks assigned to each processor at run-time.

With the partitioning algorithm, most tasks are assigned to a processor (and thereby will only execute on this processor at run-time). We call these tasks *non-split tasks*. The other tasks are called *split tasks*, since they are split into several *subtasks*. Each subtask of a split task τ_i is assigned to (and thereby executes on) a different processor, and the sum of the execution times of all subtasks equals C_i. For example, in Fig. 7.1 task τ_i is split into three subtasks τ_i^1, τ_i^2, and τ_i^3, executing on processor P_1, P_2, and P_3, respectively.

The subtasks of a task need to be synchronized to execute correctly. For example, in Fig. 7.1, τ_i^2 should not start execution until τ_i^1 is finished. This equals deferring the actual ready time of τ_i^2 by up to R_i^1 (relative to τ_i's original release time), where R_i^1 is τ_i^1's worst-case response time. One can regard this as shortening the actual relative deadline of τ_i^2 by up to R_i^1. Similarly, the actual ready time of τ_i^3 is deferred by up to $R_i^1 + R_i^2$, and τ_i^3's actual relative deadline is shortened by up to $R_i^1 + R_i^2$. We use τ_i^k to denote the kth subtask of a split task τ_i, and define τ_i^k's *synthetic deadline* as

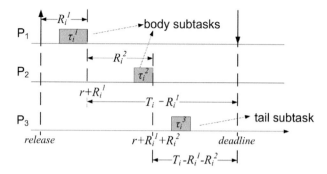

Fig. 7.1 An illustration of task splitting

$$\triangle_i^k = T_i - \sum_{l\in[1,k-1]} R_i^l. \tag{7.1}$$

Thus, we represent each subtask τ_i^k by a 3-tuple $\langle C_i^k, T_i, \triangle_i^k \rangle$, in which C_i^k is the execution time of τ_i^k, T_i is the original period, and $\triangle_i^k$ is the synthetic deadline. For consistency, each non-split task τ_i can be represented by a single subtask τ_i^1 with $C_i^1 = C_i$ and $\triangle_i^1 = T_i$. We use $U_i^k = C_i^k/T_i$ to denote a subtask τ_i^k's utilization.

We call the last subtask of τ_i its *tail subtask*, denoted by τ_i^t and the other subtasks its *body subtasks*, as shown in Fig. 7.1. We use $\tau_i^{b_j}$ to denote the jth body subtask.

We use $\tau(P_q)$ to denote the set of tasks τ_i assigned to processor P_q, and say P_q is the *host processor* of τ_i. We use $\mathscr{U}(P_q)$ to denote the sum of the utilization of all tasks in $\tau(P_q)$. A task set τ is *schedulable* under a partitioned scheduling algorithm $\mathscr{A}$, if (i) each task (subtask) has been assigned to some processor by $\mathscr{A}$'s partitioning algorithm, and (ii) each task (subtask) is guaranteed to meet its deadline under $\mathscr{A}$'s scheduling algorithm.

7.4 Parametric Utilization Bounds

On uni-processors, a *Parametric Utilization Bound* (**PUB** for short) $\Lambda(\tau)$ for a task set τ is the result of applying a function $\Lambda(\cdot)$ to τ's task parameters, such that all the tasks in τ are guaranteed to meet their deadlines on a uni-processor if τ's total utilization $\mathscr{U}(\tau) \leq \Lambda(\tau)$. We can overload this concept for multiprocessor scheduling by using τ's normalized utilization $\mathscr{U}_M(\tau)$ instead of $\mathscr{U}(\tau)$.

There have been several **PUB**s derived for **RMS** on uni-processors. The following are some examples:

- The famous *L&L* bound, denoted by $\Theta(\tau)$, is a **PUB** regarding the number of tasks N: $\Theta(\tau) = N(2^{1/N} - 1)$

- The harmonic chain bound: HC-Bound$(\tau) = K(2^{1/K} - 1)$ [45], where K is the number of harmonic chains in the task set. The 100 % bound for harmonic task sets is a special case of the harmonic chain bound with $K = 1$.
- T-Bound(τ) [163] is a PUB regarding the number of tasks and the task periods: T-Bound$(\tau) = \sum_{i=1}^{N} \frac{T'_{i+1}}{T'_i} + 2 \cdot \frac{T'_1}{T'_N} - N$, where T'_i is τ_i's *scaled period* [163].
- R-Bound(τ) [163] is similar to T-Bound(τ), but uses a more abstract parameter r, the ratio between the minimum and maximum scaled period of the task set:

$$\text{R-Bound}(\tau) = (N - 1)(r^{1/(N-1)} - 1) + 2/r - 1.$$

We observe that all the above PUBs have the following property: for any τ' obtained by decreasing the execution times of some tasks of τ, the bound $\Lambda(\tau)$ is still a valid utilization bound to guarantee the schedulability of τ'. We call a PUB holding this property a *deflatable* parametric utilization bound (called D-PUB for short).[2] We use the following lemma to precisely describe this property:

Lemma 7.1. *Let $\Lambda(\tau)$ be a D-PUB derived from the task set τ. We decrease the execution times of some tasks in τ to get a new task set τ'. If τ' satisfies $\mathscr{U}(\tau') \leq \Lambda(\tau)$, then it is guaranteed to be schedulable by RMS on a uni-processor.*

The deflatable property is very common: Actually all the PUBs we are aware of are deflatable, including the ones listed above and the non-closed-form bounds in [164]. The deflatable property is of great relevance in partitioned multiprocessor scheduling, since a task set τ will be partitioned into several subsets and each subset is executed on a processor individually. Further, due to the task splitting, a task could be divided into several subtasks, each of which holds a portion of the execution demand of the original task. So the deflatable property is clearly required to generalize a utilization bound to multiprocessors.

However, the deflatable property by itself is *not* sufficient for the generalization of a PUB $\Lambda(\tau)$ to multiprocessors. For example, suppose the harmonic task set τ in Fig. 7.2a is partitioned as in Fig. 7.2b, where τ_2 is split into τ_2^1 and τ_2^2. To correctly execute τ_2, τ_2^1 and τ_2^2 need to be synchronized such that τ_2^2 never starts execution before its predecessor τ_2^1 is finished. This can be viewed as shortening τ_2^2's relative deadline for a certain amount of time from τ_2's original deadline, as shown in Fig. 7.2c. In this case, τ_2^2 does not comply with the *L&L* task model (which requires the relative deadline to equal the period), so none of the parametric utilization bounds for the *L&L* task model are applicable to processor P_2. In [68], this problem is solved by representing τ_2^2's period by its relative deadline, as shown in Fig. 7.2d. This transforms the task set $\{\tau_1, \tau_2^2\}$ into an *L&L* task set $\{\tau_1, \tau_2^{2*}\}$, with which we

---

[2]There is a subtle difference between the *deflatable* property and the *(self-)sustainable* property [144, 147]. The deflatable property does *not* require the original task set τ to satisfy $\mathscr{U}(\tau) \leq \Lambda(\tau)$. $\mathscr{U}(\tau)$ is typically larger than 100 % since τ will be scheduled on M processors. $\Lambda(\tau)$ is merely a value obtained by applying the function $\Lambda(\cdot)$ to τ's parameters, and will be used to each individual processor.

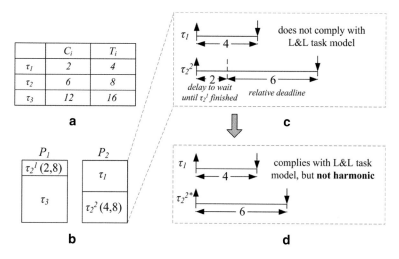

Fig. 7.2 Partitioning a harmonic task set results in a nonharmonic task set on some processor

can apply the *L&L* bound. However, this solution does not in general work for other parametric utilization bounds: In our example, we still want to apply the 100 % bound which is specific to harmonic task sets. But if we use τ_2^2's deadline 6 to represent its period, the task set $\{\tau_1, \tau_2^{2*}\}$ is not harmonic, so the 100 % bound is not applicable. This problem will be solved by our new algorithms and novel proof techniques in the following sections.

7.5   The Algorithm for Light Tasks

In the following we introduce the first algorithm **SPA1**, which achieves $\Lambda(\tau)$ (any D-PUB derived from τ's parameters), if τ is *light* in the sense of an upper bound on each task's individual utilization as follows.

Definition 7.1. A task τ_i is a *light task* if $U_i \leq \frac{\Theta(\tau)}{1+\Theta(\tau)}$, where $\Theta(\tau)$ denotes the *L&L* bound. Otherwise, τ_i is a *heavy task*. A task set τ is a *light task set* if all tasks in τ are light. $\frac{\Theta(\tau)}{1+\Theta(\tau)}$ is about 40.9 % as the number of tasks in τ grows to infinity.

For example, we can instantiate this result by the 100 % utilization bound for harmonic task sets: Let τ be any *harmonic* task set in which each task's individual utilization is no larger than 40.9 %. τ is schedulable by our algorithm **SPA1** on M processors if its normalized total utilization $\mathscr{U}_M(\tau)$ is no larger than 100 %.

7.5.1   *Algorithm Description*

The partitioning algorithm of **SPA1** is quite simple. We describe it briefly as follows:

1. Tasks are assigned in increasing priority order. We always select the processor on which the total utilization of the tasks that have been assigned so far is *minimal* among all processors.
2. A task (subtask) can be entirely assigned to the current processor, if all tasks (including the one to be assigned) on this processor can meet their deadlines under **RMS**.
3. When a task (subtask) cannot be assigned entirely to the current processor, we split it into two parts.[3] The first part is assigned to the current processor. The splitting is done such that the portion of the first part is as big as possible, guaranteeing no task on this processor misses its deadline under **RMS**; the second part is left for the assignment in the step.

Note that the difference between **SPA1** and the algorithm in [68] is that **SPA1** uses the exact RTA, instead of the utilization threshold, to determine whether a (sub)task can fit in a processor without causing deadline miss.

Algorithms 1 and 2 describe the partitioning algorithm of **SPA1** in pseudo-code. At the beginning, tasks are sorted (and will therefore be assigned) in increasing priority order, and all processors are marked as *non-full* which means they still can accept more tasks. At each step, we pick the next task in order (the one with the lowest priority), select the processor with the minimal total utilization of tasks that have been assigned so far, and invoke the routine $\mathsf{Assign}(\tau_i^k, P_q)$ to do the task assignment. $\mathsf{Assign}(\tau_i^k, P_q)$ first verifies that after assigning the task, all tasks on that processor would still be schedulable under **RMS**. This is done by applying exact schedulability analysis of calculating the response time R_j^h of each (sub)task τ_j^k on P_q after assigning this new task τ_i^k, and compare R_j^h to its (synthetic) deadline Δ_j^h. If the response time does not exceed the synthetic deadline for any of the tasks

---

1: Task order $\tau_N^1, \ldots, \tau_1^1$ by increasing priorities
2: Mark all processors as *non-full*
3: **while** exists an *non-full* processor **and** an unassigned task **do**
4:     Pick next unassigned task τ_i^k,
5:     Pick *non-full* processor P_q with minimal $\mathscr{U}(P_q)$
6:     $\mathsf{Assign}(\tau_i^k, P_q)$
7: **end while**
8: If there is an unassigned task, the algorithm **fails**,
    otherwise it **succeeds**.

Algorithm 1: The Partitioning Algorithm of **SPA1**

---

[3]In general a task may be split into more than two subtasks. Here we mean at each step the currently selected task (subtask) is split into two parts.

```
1: if τ(P_q) with τ_i^k is still schedulable then
2:     Add τ_i^k to τ(P_q)
3: else
4:     Split τ_i^k via (τ_i^k, τ_i^{k+1}) := MaxSplit(τ_i^k, P_q)
5:     Add τ_i^k to τ(P_q)
6:     Mark P_q as full
7:     τ_i^{k+1} is the next task to assign
8: end if
```

Algorithm 2: The $\mathsf{Assign}(\tau_i^k, P_q)$ Routine

on P_q, we can conclude that τ_i^k can safely be assigned to P_q without causing any deadline miss. Note that a subtask's synthetic deadline Δ_j^k may be different from its period T_j. After presenting how the overall partitioning algorithm works, we will show how to calculate Δ_j^k easily.

If τ_i^k cannot be entirely assigned to the currently selected processor P_q, it will be split into two parts using routine $\mathsf{MaxSplit}(\tau_i^k, P_q)$: the first part that makes maximum use of the selected processor, and a remaining part of that task, which will be subject to assignment in the next iteration. The desired property here is that we want the first part to be as big as possible such that, after assigning it to P_q, all tasks on that processor will still be able to meet their deadlines. In order to state the effect of $\mathsf{MaxSplit}(\tau_i^k, P_q)$ formally, we introduce the concept of a *bottleneck*:

Definition 7.2. A *bottleneck* of processor P_q is a (sub)task that is assigned to P_q, and will become unschedulable if we increase the execution time of the task with the highest priority on P_q by an arbitrarily small positive number.

Note that there may be more than one bottleneck on a processor. Further, since SPA1 assigns tasks in increasing priority order, MaxSplit always operates on the task that has the highest priority on the processor in question. So we can state:

Definition 7.3. $\mathsf{MaxSplit}(\tau_i^k, P_q)$ is a function that splits τ_i^k into two subtasks τ_i^k and τ_i^{k+1} such that

1. τ_i^k can now be assigned to P_q without making any task in $\tau(P_q)$ unschedulable.
2. After assigning τ_i^k, P_q has a bottleneck.

MaxSplit can be implemented by, for example, performing a binary search over $[0, C_i^k]$ to find out the maximal portion of τ_i^k with which all tasks on P_q can meet their deadlines. A more efficient implementation of MaxSplit was presented in [151], in which one only needs to check a (small) number of possible values in $[0, C_i^k]$. The complexity of this improved implementation is still pseudo-polynomial, but in practice it is very efficient.

The while loop in SPA1 terminates as soon as all processors are "full" *or* all tasks have been assigned. If the loop terminates due to the first reason and there are still unassigned tasks left, the algorithm reports a failure of the partitioning, otherwise a success.

Calculating Synthetic Deadlines

Now we show how to calculate each (sub)task τ_i^k's synthetic deadline Δ_i^k, which was left open in the above presentation. If τ_i^k is a non-split task, its synthetic deadline trivially equals its period T_i.

We consider the case that τ_i^k is a split subtask. Since tasks are assigned in increasing order of priorities, and a processor is *full* after a body subtask is assigned to it, we have the following lemma:

Lemma 7.2.   *A body subtask has the highest priority on its host processor.*

A consequence is that, the response time of each body subtask equals its execution time, and one can replace R_i^l by C_i^l in (7.1) to calculate the synthetic deadline of a subtask. Especially, we are interested in the synthetic deadlines of tail subtasks (we don't need to worry about a body subtask's synthetic deadline since it has the highest priority on its host processor and is schedulable anyway). The calculation is stated in the following lemma.

Lemma 7.3.   *A tail subtask τ_i^t's synthetic deadline Δ_i^t is calculated by*

$$\Delta_i^t = T_i - C_i^{body}$$

where C_i^{body} is the execution time sum of τ_i's body subtasks.

Scheduling at Run-Time

At run-time, the tasks will be scheduled according to the **RMS** priority order on each processor locally, i.e., with their original priorities. The subtasks of a split task respect their precedence relations, i.e., a split subtask τ_i^k is ready for execution when its preceding subtask τ_i^{k-1} on some other processor has finished.

From the presented partitioning and scheduling algorithm of **SPA1**, it is clear that successful partitioning implies schedulability (remember that for split tasks, the synchronization delays have been counted into the synthetic deadlines, which are the ones used in the RTA to determine whether a task is schedulable). We state this in the following lemma:

Lemma 7.4.   *Any task set that has been successfully partitioned by* **SPA1** *is schedulable.*

7.5.2   Utilization Bound

We will now prove that SPA1 has the utilization bound of $\Lambda(\tau)$ for *light* task sets, i.e., if a light task set τ is not successfully partitioned by SPA1, then the sum of the assigned utilizations of all processors is *at least*[4] $M \cdot \Lambda(\tau)$.

In order to show this, we assume that the assigned utilization on some processor is *strictly less* than $\Lambda(\tau)$. We prove that this implies there is no bottleneck on that processor. This is a contradiction, because each processor with which MaxSplit has been used must have a bottleneck. We also know that MaxSplit was used for all processors, since the partitioning failed.

In the following, we assume P_q to be a processor with an assigned utilization of $U(P_q) < \Lambda(\tau)$. A task on P_q is either a non-split task, a body subtask or a tail subtask. The main part of the proof consists of showing that P_q cannot have a bottleneck of any type.

As the first step, we show this for non-split tasks and body subtasks (Lemma 7.5), after which we deal with the more difficult case of tail subtasks (Lemma 7.7).

Lemma 7.5. *Suppose task set τ is not schedulable by* SPA1, *and after the partitioning phase it holds for a processor P_q that*

$$\mathcal{U}(P_q) < \Lambda(\tau) \tag{7.2}$$

Then a bottleneck of P_q is neither a non-split task nor a body subtask.

Proof. By Lemma 7.2 we know that the body subtask has the highest priority on P_q, so it can never be a bottleneck.

For the case of non-split tasks, we will show that Condition (7.2) is sufficient for their deadlines to be met. The key observation is that although some split tasks on this processor may have a shorter deadline than period, this does not change the scheduling behavior of RMS, so $\Lambda(\tau)$ is still sufficient to guarantee the schedulability of a non-split task. For a more precise proof, we use Γ to denote the set of tasks on P_q, and construct a new task set Γ^* corresponding to Γ such that each non-split task τ_i in Γ has a counterpart in Γ^* that is exactly the same as τ_i, and each split subtask in Γ has a counterpart in Γ^* with deadline changed to equal its period. It's easy to see that Γ^* can be obtained by decreasing some tasks' execution times in the original task set τ (a task in τ but not Γ^* can be considered as the case that we decrease its execution time to 0). By Lemma 7.1 and Condition (7.2) we know, the deflatable utilization bound $\Lambda(\tau)$ guarantee Γ^*'s schedulability. Thus, if the execution time of the highest-priority task on P_q is increased by an arbitrarily small amount ε such that the total utilization still does not exceed $\Lambda(\tau)$, Γ^* will still be schedulable. Recall that the only difference between Γ and Γ^* is the subtasks' deadlines, and since the scheduling behavior of RMS does not depend

---

[4]By this, the normalized utilization of τ *strictly exceeds* $\Lambda(\tau)$, since there are (sub)tasks not assigned to any of the processors after a failed partitioning.

on task deadlines (remember that at this moment we only want to guarantee the schedulability of non-split tasks), we can conclude that each non-split task in Γ is also schedulable, which is still true after increasing ε to the highest priority task on P_q.

In the following we prove that in a light task set, a bottleneck on a processor with utilization lower than $\Lambda(\tau)$ is not a tail subtask either. The proof goes in two steps: We first derive in Lemma 7.6 a general condition guaranteeing that a tail subtask cannot be a bottleneck; then, we conclude in Lemma 7.7 that a bottleneck on a processor with utilization lower than $\Lambda(\tau)$ is not a tail subtask, by showing that the condition in Lemma 7.6 holds for each of these tail subtasks.

We use the following notation: Let τ_i be a task split into B body subtasks $\tau_i^{b_1} \dots \tau_i^{b_B}$, assigned to processors $P_{b_1} \dots P_{b_B}$, respectively, and a tail subtask τ_i^t assigned to processor P_t. The utilization of the tail subtask τ_i^t is $U_i^t = \frac{C_i^t}{T_i}$, and the utilization of a body subtask $\tau_i^{b_j}$ is $U_i^{b_j} = \frac{C_i^{b_j}}{T_i}$. We use U_i^{body} to denote the total utilization of τ_i's all body subtasks:

$$U_i^{body} = \sum_{j \in [1,B]} U_i^{b_j} = U_i - U_i^t$$

For the tail subtask τ_i^t, let X_t denote the total utilization of all (sub)tasks assigned to P_t with *lower* priority than τ_i^t, and Y_t the total utilization of all (sub)tasks assigned to P_t with *higher* priority than τ_i^t.

For each body subtask $\tau_i^{b_j}$, let X_{b_j} denote the total utilization of all (sub)tasks assigned to P_{b_j} with *lower* priority than $\tau_i^{b_j}$. (We do not need Y_{b_j}, since by Lemma 7.2 we know no task on P_{b_j} has higher priority than τ_i.)

We start with the general condition identifying non-bottleneck tail subtasks.

Lemma 7.6. *Suppose a tail subtask τ_i^t is assigned to processor P_t and $\Theta(\tau)$ is the L&L bound. If*

$$Y_t + U_i^t < \Theta(\tau) \cdot (1 - U_i^{body}) \tag{7.3}$$

then τ_i^t is not a bottleneck of processor P_t.

Proof. The lemma is proved by showing τ_i^t is still schedulable after increasing the utilization of the task with the highest priority on P_t by a small number ϵ such that : $(Y_t + \epsilon) + U_i^t < \Theta(\tau) \cdot (1 - U_i^{body})$ (note that one can always find such an ϵ). By the definition of U_i^{body} and $\triangle_i^t$, this equals

$$((Y_t + \epsilon) + U_i^t) \cdot T_i / \triangle_i^t < \Theta(\tau) \tag{7.4}$$

The key of the proof is to show that Condition (7.4) still guarantees that τ_i^t can meet its deadline. Note that one cannot directly apply the L&L bound $\Theta(\tau)$ to the task set Γ consisting of τ_i^t and the tasks contributing to Y_t, since τ_i^t's deadline is shorter than its period, i.e., Γ does not comply with the *L&L* task model. In our proof, this problem is solved by the "period shrinking" technique [68]: we transform Γ into an *L&L* task set Γ^* by reducing some of the task periods, and prove that the total utilization of Γ^* is bounded by the LHS of (7.4), and thereby bounded by $\Theta(\tau)$. On the other hand, the construction of Γ^* guarantees that the schedulability of Γ^* implies the schedulability of τ_i^t. See [68] for details about the "period shrinking" technique.

Note that in Condition (7.3) of Lemma 7.6, the *L&L* bound $\Theta(\tau)$ is involved. This is because in its proof we need to use the *L&L* bound $\Theta(\tau)$, rather than the higher parametric bound $\Lambda(\tau)$, to guarantee the schedulability of the constructed task set Γ^* where some task periods are decreased. For example, suppose the original task set is harmonic, the constructed set Γ^* may not be harmonic since some of task periods are shortened to $\triangle_i^t$, which is not necessarily harmonic with other periods. So the 100 % bound of harmonic task sets does not apply to Γ^*. However, $\Theta(\tau)$ is still applicable, since it only depends on, and is monotonically decreasing with respect to the task number.

Having this lemma, we now show that a tail subtask τ_i^t cannot be a bottleneck either, if its host processor's utilization is less than $\Lambda(\tau)$, by proving Condition (7.3) for τ_i^t.

Lemma 7.7. *Let τ be a* light *task set unschedulable by* SPA1, *and let τ_i be a split task whose tail subtask τ_i^t is assigned to processor P_t. If*

$$\mathscr{U}(P_t) < \Lambda(\tau) \tag{7.5}$$

then τ_i^t is not a bottleneck of P_t.

Proof. The proof is by contradiction. We assume the lemma does *not* hold for one or more tasks, and let τ_i be the lowest-priority one among these tasks, i.e., τ_i^t is a bottleneck of its host processor P_t, and all tail subtasks with lower priorities are either not a bottleneck or on a processor with assigned utilization at least $\Lambda(\tau)$.

Recall that $\{\tau_i^{b_j}\}_{j\in[1,B]}$ are the body subtasks of τ_i, and P_t and $\{P_{b_j}\}_{j\in[1,B]}$ are processors hosting the corresponding tail and body subtasks. Since a body task has the highest priority on its host processor (Lemma 7.3) and tasks are assigned in increasing priority order, all tail subtasks on processors $\{P_{b_j}\}_{j\in[1,B]}$ have lower priorities than τ_i.

We will first show that all processors $\{P_{b_j}\}_{j\in[1,B]}$ have an individual assigned utilization at least $\Lambda(\tau)$. We do this by contradiction: Assume there is a P_{b_j} with $\mathscr{U}(P_{b_j}) < \Lambda(\tau)$. Since tasks are assigned in increasing priority order, we know any tail subtask on P_{b_j} has lower priority than τ_i. And since τ_i is the lowest-priority task violating the lemma and $\mathscr{U}(P_{b_j}) < \Lambda(\tau)$, we know any tail subtask on P_{b_j} is not a bottleneck. At the same time, $\mathscr{U}(P_{b_j}) < \Lambda(\tau)$ also implies the non-split tasks and body subtasks on P_{b_j} are not bottlenecks either (by Lemma 7.5). So we can conclude that there is no bottleneck on P_{b_j} which contradicts the fact there is at least one bottleneck on each processor. So the assumption of P_{b_j}'s assigned utilization being lower than $\Lambda(\tau)$ must be false, by which we can conclude that all processors hosting τ_i^t's body tasks have assigned utilization at least $\Lambda(\tau)$. Thus we have

$$\sum_{j\in[1,B]} \underbrace{(U_i^{b_j} + X_{b_j})}_{\mathscr{U}(P_{b_j})} \geq B \cdot \Lambda(\tau) \tag{7.6}$$

Further, the assumption from Condition (7.5) can be rewritten as

$$X_t + Y_t + U_i^t < \Lambda(\tau) \tag{7.7}$$

We combine (7.6) and (7.7) into

$$X_t + Y_t + U_i^t < \frac{1}{B} \sum_{j\in[1,B]} (U_i^{b_j} + X_{b_j})$$

Since the partitioning algorithm selects at each step the processor on which the so-far assigned utilization is minimal, we have $\forall j \in [1, B] : X_{b_j} \leq X_t$. Thus, the inequality can be relaxed to:

$$Y_t + U_i^t < \frac{1}{B} \sum_{j\in[1,B]} U_i^{b_j}$$

We also have $B \geq 1$ and $U_i^{body} = \sum_{j\in[1,B]} U_i^{b_j}$, so

$$Y_t + U_i^t < U_i^{body}$$

Now, in order to get to Condition (7.3), which implies τ_i^t is not a bottleneck (Lemma 7.6), we need to show that the RHS of this inequality is bounded by the RHS of Condition (7.3), i.e.,

$$U_i^{body} \leq \Theta(\tau)(1 - U_i^{body})$$

It is easy to see that this is equivalent to the following, which holds since τ_i is by assumption a light task:

$$U_i^{body} \leq \frac{\Theta(\tau)}{1 + \Theta(\tau)}$$

By now we have proved Condition (7.3) for τ_i^t and by Lemma 7.6 we know τ_i^t is not a bottleneck on P_t, which contradicts to our assumption.

We are ready to present **SPA1**'s utilization bound.

Theorem 7.1. $\Lambda(\tau)$ *is a utilization bound of* SPA1 *for light task sets, i.e., any light task set* τ *with*

$$\mathscr{U}_M(\tau) \leq \Lambda(\tau)$$

is schedulable by SPA1*.*

Proof. Assume a light task set τ with $\mathscr{U}_M(\tau) \leq \Lambda(\tau)$ is not schedulable by SPA1, i.e., there are tasks not assigned to any of the processors after the partitioning procedure with τ. By this we know the sum of the assigned utilization of all processors after the partitioning is *strictly less* than $M \cdot \Lambda(\tau)$, so there is at least one processor P_q with a utilization *strictly less* than $\Lambda(\tau)$. By Lemma 7.5 we know the bottleneck of this processor is neither a non-split task nor a body subtask, and by Lemma 7.7 we know the bottleneck is not a tail subtask either, so there is no bottleneck on this processor. This contradicts the property of the partitioning algorithm that all processors to which no more task can be assigned must have a bottleneck.

7.6   The Algorithm for Any Task Set

In this section, we introduce SPA2, which removes the restriction to light task sets in SPA1. We will show that SPA2 can achieve a D-PUB $\Lambda(\tau)$ for any task set τ, if $\Lambda(\tau)$ does not exceed $\frac{2\Theta(\tau)}{1+\Theta(\tau)}$. In other words, if one can derive a D-PUB $\Lambda'(\tau)$ from τ's parameters under uni-processor RMS, SPA2 can achieve the utilization bound of $\Lambda(\tau) = \min(\Lambda'(\tau), \frac{2\Theta(\tau)}{1+\Theta(\tau)})$. Note that $\frac{2\Theta(\tau)}{1+\Theta(\tau)}$ is decreasing with respect to N, and it is around 81.8 % when N goes to infinity. For example, we can instantiate our result by the harmonic chain bound $K(2^{1/K} - 1)$:

- $K = 3$. Since $3(2^{1/3} - 1) \approx 77.9 \% < 81.8 \%$, we know that *any* task set τ in which there are at most 3 harmonic chains is schedulable by our algorithm SPA2 on M processors if its normalized utilization $\mathscr{U}_M(\tau)$ is no larger than 77.9 %.
- $K = 2$. Since $2(2^{1/2} - 1) \approx 82.8 \% > 81.8 \%$, we know 81.8 % can be used as the utilization bound in this case: *any* task set τ in which there are at most 2 harmonic chains is schedulable by our algorithm SPA2 on M processors if its normalized utilization $\mathscr{U}_M(\tau)$ is no larger than 81.8 %.

So we can see that despite an upper bound on $\Lambda(\tau)$, SPA2 still provides significant room for higher utilization bounds.

For simplicity of presentation, we assume each task's utilization is bounded by $\Lambda(\tau)$. Note that this assumption does not invalidate the utilization bound of our algorithm for task sets which have some individual task's utilization above $\Lambda(\tau)$.[5]

SPA2 adds a pre-assignment mechanism to handle the heavy tasks. In the pre-assignment, we first identify the heavy tasks whose tail subtasks would have low priority if they were split, and pre-assign these tasks to one processor each, which avoids the split. The identification is checked by a simple test condition, called *pre-assign condition*. Those heavy tasks that do not satisfy this condition will be assigned (and possibly split) later, together with the light tasks. Note that the number of tasks need to be pre-assigned is at most the number of processors. This will be clear in the algorithm description.

We introduce some notations. If a heavy task τ_i is pre-assigned to a processor P_q, we call τ_i a *pre-assigned task* and P_q a *pre-assigned processor*, otherwise τ_i a *normal task* and P_q a *normal processor*.

7.6.1 Algorithm Description

The partitioning algorithm of **SPA2** contains three phases:

1. We first pre-assign the heavy tasks that satisfy the *pre-assign condition* to one processor each, in decreasing priority order.
2. We do task partitioning with the remaining (i.e., normal) tasks and remaining (i.e., normal) processors similar to **SPA1** until all the normal processors are full.
3. The remaining tasks are assigned to the pre-assigned processors in increasing priority order; the assignment selects the processor hosting the lowest-priority pre-assigned task, to assign as many tasks as possible until it is full, then selects the next processor.

The pseudo-code of **SPA2** is given in Algorithm 3. At the beginning of the algorithm, all the processors are marked as *normal* and *non-full*. In the first phase, we visit all the tasks in decreasing priority order, and for each *heavy* task we determine whether we should pre-assign it or not, by checking the *pre-assign condition*:

$$\sum_{i<j} U_j \leq (|\mathscr{P}^{\triangleright}(\tau_i)| - 1) \cdot \Lambda(\tau) \tag{7.8}$$

where $|\mathscr{P}^{\triangleright}(\tau_i)|$ is the number of processors marked as *normal* at the moment we are checking for τ_i. If this condition is satisfied, we pre-assign this heavy task to

---

[5]One can let tasks with a utilization more than $\Lambda(\tau)$ execute exclusively on a dedicated processor each. If we can prove that the utilization bound of all the other tasks on all the other processors is $\Lambda(\tau)$, then the utilization bound of the overall system is also at least $\Lambda(\tau)$.

1: Mark all processors as *normal* and *non-full*

   // *Phase 1: Pre-assignment*
2: Sort all tasks in τ in *decreasing* priority order
3: **for** each task in τ **do**
4:     Pick next task τ_i
5:     **if** DeterminePreAssign(τ_i) **then**
6:         Pick the *normal* processor with the minimal index P_q
7:         Add τ_i to $\tau(P_q)$
8:         Mark P_q as *pre-assigned*
9:     **end if**
10: **end for**

   // *Phase 2: Assign remaining tasks to normal processors*
11: Sort all unassigned tasks in *increasing* priority order
12: **while** there is a *non-full normal* processor
            and an unassigned task **do**
13:     Pick next unassigned task τ_i
14:     Pick the *non-full normal* processor P_q with minimal $\mathcal{U}(P_q)$
15:     $Assign(\tau_i^k, P_q)$
16: **end while**

   // *Phase 3: Assign remaining tasks to pre-assigned processors*
   // *Remaining tasks are still in increasing priority order*
17: **while** there is a *non-full pre-assigned* processor
            and an unassigned task **do**
18:     Pick next unassigned task τ_i
19:     Pick the *non-full pre-assigned* processor P_q with the
        largest index
20:     $Assign(\tau_i^k, P_q)$
21: **end while**

22: If there is an unassigned task, the algorithm **fails**,
    otherwise it **succeeds**.

Algorithm 3: The Partitioning Algorithm of SPA2

the current selected processor, which is the one with the minimal index among all normal processors, and mark this processor as *pre-assigned*. Otherwise, we do not pre-assign this heavy task, and leave it to the following phases. The intuition of the pre-assign condition (7.8) is: We pre-assign a heavy task τ_i if the total utilization of lower-priority tasks is relatively small, since otherwise its tail subtask may end up with a low priority on the corresponding processor. Note that, no matter how many heavy tasks are there in the system, the number of pre-assigned tasks is at most the number of processors: after $|\mathscr{P}^{\triangleright}(\tau_i)|$ reaching 0, the pre-assign condition never holds, and no more heavy task will be pre-assigned.

In the second phase we assign the remaining tasks to *normal* processors only. Note that the remaining tasks are either light tasks or the heavy tasks that do not satisfy the pre-assign condition. The assignment policy in this phase is the same as for SPA1: We sort tasks in increasing priority order, and at each step select the

```
1:  𝒫▷(τᵢ) := the set of normal processors at this moment
2:  if τᵢ is heavy then
3:     if ∑_{j>i} Uⱼ ≤ (|𝒫▷(τᵢ)| − 1) · Λ(τ) then
4:        return true
5:     end if
6:  end if
7:  return false
```

Algorithm 4: The DeterminePreAssign(τ_i) Routine

normal processor P_q with the minimal assigned utilization. Then we do the task assignment: we either add τ_i^k to $\tau(P_q)$ if τ_i^k can be entirely assigned to P_q, or split τ_i^k and assigns a maximized portion of it to P_q otherwise.

In the third phase we continue to assign the remaining tasks to *pre-assigned* processors. There is an important difference between the second phase and the third phase: In the second phase tasks are assigned by a "worst-fit" strategy, i.e., the utilization of all processors is increased "evenly," while in the third phase tasks are now assigned by a "first-fit" strategy. More precisely, we select the pre-assigned processor which hosts the lowest-priority pre-assigned task of all non-full processors. We assign as much workload as possible to it, until it is full, and then move to the next processor. This strategy is one of the key points to facilitate the induction-based proof of the utilization bound in the next subsection.

After these three phases, the partitioning fails if there still are unassigned tasks left, otherwise it is successful. At run-time, the tasks assigned to each processor are scheduled by RMS with their original priorities, and the subtasks of a split task need to respect their precedence relations, which is the same as in SPA1.

Note that, when Assign calculates the synthetic deadlines and verifies whether the tasks assigned to a processor are schedulable, it assumes that any body subtask has the highest priority on its host processor, which has been proved true for SPA1 in Lemma 7.2. It is easy to see that this assumption also holds for the second phase of SPA2 (the task assignment on normal processors), in which tasks are assigned in exactly the same way as SPA1. But it is not clear for this moment whether this assumption also holds for the third phase or not, since there are pre-assigned tasks already assigned to these pre-assigned processors in the first phase, and there is a risk that a pre-assigned task might have higher priority than the body subtask on that processor. However, as will be shown in the proof of Lemma 7.13, *a body subtask on a pre-assigned processor has the highest priority on its host processor*, thus routine Assign indeed performs a correct schedulability analysis for task assignment and splitting, by which we know any task set successfully partitioned by SPA2 is guaranteed to meet all deadlines at run-time.

7.6.2 Utilization Bound

The proof of the utilization bound $\Lambda(\tau)$ for SPA2. follows a similar pattern as the proof for SPA1, by assuming a task set τ that can't be completely assigned. The main difficulty is that we now have to deal with heavy tasks as well. Recall that the approach in Sect. 7.5 was to show an individual utilization of at least $\Lambda(\tau)$ *on each single processor* after an "overflowed" partitioning phase. However, for SPA2, we will not do that directly. Instead, we will show the appropriate bound for *sets of processors*.

We first introduce some additional notation. Let's assume that $K \geq 0$ heavy tasks are pre-assigned in the first phase of SPA2. Then $\mathscr{P}$ is partitioned into the set of *pre-assigned* processors:

$$\mathscr{P}^{\mathscr{P}} := \{P_1, \ldots, P_K\}$$

and the set of *normal* processors:

$$\mathscr{P}^{\mathscr{N}} := \{P_{K+1}, \ldots, P_M\}.$$

We also use

$$\mathscr{P}_{\geq q} := \{P_q, \ldots, P_M\}$$

to denote the set of processors with index of at least q.

We want to show that, after a failed partitioning procedure of τ, the total utilization sum of all processors is at least $M \cdot \Lambda(\tau)$. We do this by proving the property

$$\sum_{P_j \in \mathscr{P}_{\geq q}} \mathscr{U}(P_j) \geq |\mathscr{P}_{\geq q}| \cdot \Lambda(\tau)$$

by induction on $\mathscr{P}_{\geq q}$ for all $q \leq K$, starting with base case $q = K$, and using the inductive hypothesis with $q = m+1$ to derive this property for $q = m$. When $q = 1$, it implies the expected bound $M \cdot \Lambda(\tau)$ for all the M processors.

7.6.2.1 Base Case

The proof strategy of the base case is: We assume that the total assigned utilization of normal processors is below the expected bound, by which we can derive the absence of bottlenecks on some processors in $\mathscr{P}^{\mathscr{N}}$. This contradicts the fact that there is at least one bottleneck on each processor after a failed partitioning procedure.

First, Lemma 7.5 still holds for normal processors under SPA2, i.e., a bottleneck on a normal processor with assigned utilization lower than $\Lambda(\tau)$ is neither a non-split task nor a body subtask. This is because the partitioning procedure of SPA2 on normal processors is exactly the same as SPA1 and one can reuse the reasoning for Lemma 7.5 here. In the following, we focus on the difficult case of tail subtasks.

Lemma 7.8. *Suppose there are remaining tasks after the second phase of* SPA2. *Let τ_i^t be a tail subtask assigned to P_t. If both the following conditions are satisfied*

$$\sum_{P_q \in \mathscr{P}^{\mathcal{N}}} \mathscr{U}(P_q) < |\mathscr{P}^{\mathcal{N}}| \cdot \Lambda(\tau) \tag{7.9}$$

$$\mathscr{U}(P_t) < \Lambda(\tau) \tag{7.10}$$

then τ_i^t is not a bottleneck on P_t.

Proof. We prove by contradiction: We assume the lemma does *not* hold for one or more tasks, and let τ_i be the lowest-priority one among these tasks.

Similar with the proof of its counterpart in SPA1 (Lemma 7.7), we will first show that all processors hosting τ_i's body subtasks have assigned utilization at least $\Lambda(\tau)$. We do this by contradiction. We assume $\mathscr{U}(P_{b_j}) < \Lambda(\tau)$, and by Condition (7.9) we know the tail subtasks on P_{b_j} are not bottlenecks (the tail subtasks on P_{b_j} all satisfy this lemma, since they all have lower priorities than τ_i, and by assumption τ_i is the lowest-priority task does not satisfy this lemma). By Lemma 7.5 (which still holds for normal processors as discussed above), we know a bottleneck of P_{b_j} is neither a non-split task nor a body subtask. So we can conclude that there is no bottleneck on P_{b_j}, which is a contradiction. Therefore, we have proved that all processors hosting τ_i's body subtasks have assigned utilization at least $\Lambda(\tau)$. These results will be used later in this proof.

In the following we will prove τ_i^t is not a bottleneck, by deriving Condition (7.3) and apply Lemma 7.6 to τ_i^t. τ_i is either light or heavy. For the case τ_i is light, the proof is exactly the same as for Lemma 7.7, since the second phase of SPA2 works in exactly the same way as SPA1. Note that to prove for the light task case, only Condition (7.9) is needed (the same as in Lemma 7.7).

In the following we consider the case that τ_i is heavy. We prove in two cases:

- $U_i^{body} \geq \frac{\Lambda(\tau) - \Theta(\tau)}{1 - \Theta(\tau)}$

  Since τ_i is a heavy task but not pre-assigned, it failed the pre-assign condition, satisfying the negation of that condition:

$$\sum_{j>i} U_j > (|\mathscr{P}^{\triangleright}(\tau_i)| - 1) \cdot \Lambda(\tau) \tag{7.11}$$

We split the utilization sum of all lower-priority tasks in two parts: $\mathscr{U}^{\alpha}$, the part contributed by pre-assigned tasks, and $\mathscr{U}^{\beta}$, the part contributed by normal tasks. By the partitioning algorithm construction, we know the $\mathscr{U}^{\beta}$ part is on normal processors and the $\mathscr{U}^\alpha$ part is on processors in $\mathscr{P}^\triangleright(\tau_i) \setminus \mathscr{P}^\mathcal{N}$. We further know that each pre-assigned processor has one pre-assigned task, and each task has a utilization of at most $\Lambda(\tau)$ (our assumption stated in the beginning of Sect. 7.6). Thus, we have

$$\mathscr{U}^\beta \leq (|\mathscr{P}^\triangleright(\tau_i)| - |\mathscr{P}^\mathcal{N}|) \cdot \Lambda(\tau) \tag{7.12}$$

By replacing $\sum_{j>i} U_j$ by $\mathscr{U}^\alpha + \mathscr{U}^\beta$ in (7.11) and applying (7.12), we get

$$\mathscr{U}^\alpha > (|\mathscr{P}^\mathcal{N}| - 1) \cdot \Lambda(\tau) \tag{7.13}$$

The assigned utilizations on processors in $\mathscr{P}^\mathcal{N}$ consists of three parts: (i) the utilization of tasks with lower priority than τ_i, (ii) the utilization of τ_i, and (iii) the utilization of tasks with higher priority than τ_i. We know that part (i) is $\mathscr{U}^\alpha$, part (ii) is U_i, and the part (iii) is at least Y_t. So we have

$$\mathscr{U}^\alpha + U_i + Y_t \leq \sum_{P_q \in \mathscr{P}^\mathcal{N}} \mathscr{U}(P_q) \tag{7.14}$$

By Condition (7.9), (7.13), and (7.14) we get

$$U_i + Y_t \leq \Lambda(\tau)$$

In order to use this to derive Condition (7.3) of Lemma 7.6, which indicates τ_i^t is not a bottleneck, we need to prove

$$\Lambda(\tau) - U_i^{body} \leq \Theta(\tau)(1 - U_i^{body})$$

$$\Leftrightarrow U_i^{body} \geq \frac{\Lambda(\tau) - \Theta(\tau)}{1 - \Theta(\tau)} \quad (\text{since } \Theta(\tau) < 1)$$

which is obviously true by the precondition of this case.

- $U_i^{body} < \frac{\Lambda(\tau) - \Theta(\tau)}{1 - \Theta(\tau)}$

First, Condition (7.10) can be rewritten as

$$X_t + Y_t + U_i^t < \Lambda(\tau) \tag{7.15}$$

Since all processors hosting τ_i's body subtasks have assigned utilization at least $\Lambda(\tau)$ (proved in above), we have

$$\sum_{j \in [1, B_i]} X_{b_j} + U_i^{body} > B_i \cdot \Lambda(\tau)$$

Since at each step of the second phase, SPA2 always selects the processor with the minimal assigned utilization to assign the current (sub)task, we have $X_t \geq X_{b_j}$ for each X_{b_j}. Therefore we have

$$B_i X_t + U_i^{body} \geq B_i \cdot \Lambda(\tau)$$

$$\Rightarrow X_t \geq \Lambda(\tau) - U_i^{body} \quad (\text{since } B_i \geq 1)$$

combining which and (7.15) we get

$$Y_t + U_i^t < U_i^{body}$$

Now, to prove Condition (7.3) of Lemma 7.6, which indicates τ_i^t is not a bottleneck, we only need to show

$$U_i^{body} \leq \Theta(\tau)(1 - U_i^{body})$$

$$\Leftrightarrow U_i^{body} \leq \frac{\Theta(\tau)}{1 + \Theta(\tau)}$$

Due to the precondition of this case $U_i^{body} < \frac{\Lambda(\tau) - \Theta(\tau)}{1 - \Theta(\tau)}$, we only need to prove

$$\frac{\Lambda(\tau) - \Theta(\tau)}{1 - \Theta(\tau)} \leq \frac{\Theta(\tau)}{1 + \Theta(\tau)}$$

$$\Leftrightarrow \Lambda(\tau) \leq \frac{2\Theta(\tau)}{1 + \Theta(\tau)}$$

which is true since $\Lambda(\tau)$ is assumed to be at most $\frac{2\Theta(\tau)}{1+\Theta(\tau)}$ in SPA2.

In summary, we know τ_i^t is not a bottleneck.

By the above reasoning, we can establish the base case:

Lemma 7.9. *Suppose there are remaining tasks after the second phase of* SPA2 *(there exists at least one bottleneck on each normal processor). We have*

$$\sum_{P_q \in \mathscr{P}^{\mathscr{N}}} \mathscr{U}(P_q) \geq |\mathscr{P}^{\mathscr{N}}| \cdot \Lambda(\tau)$$

7.6.2.2   Inductive Step

We start with a useful property concerning the pre-assigned tasks' local priorities.

Lemma 7.10. *Suppose P_m is a pre-assigned processor. If*

$$\sum_{P_q \in \mathscr{P}_{\geq m+1}} \mathscr{U}(P_q) \geq |\mathscr{P}_{\geq m+1}| \cdot \Lambda(\tau) \tag{7.16}$$

then the pre-assigned task on P_m has the lowest *priority among all tasks assigned to P_m.*

Proof. Let τ_i be the pre-assigned task on P_m. Since τ_i is pre-assigned, we know that it satisfies the pre-assign condition:

$$\sum_{j>i} U_j \leq \underbrace{(|\mathscr{P}^{\triangleright}(\tau_i)| - 1)}_{|\mathscr{P}_{\geq m+1}|} \cdot \Lambda(\tau)$$

Using this with (7.16) we have

$$\sum_{P_q \in \mathscr{P}_{\geq m+1}} \mathscr{U}(P_q) \geq \sum_{j>i} U_j \tag{7.17}$$

which means the total capacity of the processors with larger indices is enough to accommodate all lower-priority tasks.

By the partitioning algorithm, we know that no tasks, except τ_i which has been pre-assigned already, will be assigned to P_m before all processors with larger indices are full. So no task with priority lower than τ_i will be assigned to P_m.

Now we start the main proof of the inductive step.

Lemma 7.11. *We use* **SPA2** *to partition task set τ. Suppose there are remaining tasks after processor P_m is full (there exists at least one bottleneck on P_m). If*

$$\sum_{P_q \in \mathscr{P}_{\geq m+1}} \mathscr{U}(P_q) \geq |\mathscr{P}_{\geq m+1}| \cdot \Lambda(\tau) \tag{7.18}$$

then we have

$$\sum_{P_q \in \mathscr{P}_{\geq m}} \mathscr{U}(P_q) \geq |\mathscr{P}_{\geq m}| \cdot \Lambda(\tau)$$

Proof. We prove by contradiction. Assume

$$\sum_{P_q \in \mathscr{P}_{\geq m}} \mathscr{U}(P_q) < |\mathscr{P}_{\geq m}| \cdot \Lambda(\tau) \tag{7.19}$$

With assumption (7.18) this implies the bound on P_m's utilization:

$$\mathscr{U}(P_m) < \Lambda(\tau) \tag{7.20}$$

As before, with (7.20) we want to prove that a bottleneck on P_m is neither a non-split task, a body subtask nor a tail subtask, which forms a contradiction and completes the proof. In the following we consider each type individually.

We first consider non-split tasks. Again, $\Lambda(\tau)$ is sufficient to guarantee the schedulability of non-split tasks, although the relative deadlines of split subtasks on this processor may change. Thus, (7.20) implies that a non-split task cannot be a bottleneck of P_m.

Then we consider body subtasks. By Lemma 7.10 we know the pre-assigned task has the lowest priority on P_m. We also know that all normal tasks on P_m have lower priority than the body subtask, since in the third phase of **SPA2** tasks are assigned in increasing priority order. Therefore, we can conclude that the body subtask has the highest priority on P_m, and cannot be a bottleneck.

At last we consider tail subtasks. Let τ_i^t be a tail subtask assigned to P_m. We distinguish the following two cases:

- $U_i^{body} < \frac{\Theta(\tau)}{1+\Theta(\tau)}$

  The inductive hypothesis (7.18) guarantees with Lemma 7.10 that the pre-assigned task has the lowest priority on P_m, so X_t contains at least the utilization of this pre-assigned task, which is heavy. So we have

$$X_t \geq \frac{\Theta(\tau)}{1 + \Theta(\tau)} \tag{7.21}$$

We can rewrite (7.20) as $X_t + Y_t + U_i^t < \Lambda(\tau)$ and apply it to (7.21) to get:

$$Y_t + U_i^t < \Lambda(\tau) - \frac{\Theta(\tau)}{1 + \Theta(\tau)} \tag{7.22}$$

Recall that $\Lambda(\tau)$ is restricted by an upper bound in **SPA2**:

$$\Lambda(\tau) \leq \frac{2\Theta(\tau)}{1 + \Theta(\tau)}$$

$$\Leftrightarrow \Lambda(\tau) - \frac{\Theta(\tau)}{1 + \Theta(\tau)} \leq \Theta(\tau)(1 - \frac{\Theta(\tau)}{1 + \Theta(\tau)})$$

By applying $U_i^{body} < \frac{\Theta(\tau)}{1+\Theta(\tau)}$ to above we have

$$\Lambda(\tau) - \frac{\Theta(\tau)}{1 + \Theta(\tau)} < \Theta(\tau)(1 - U_i^{body})$$

And by (7.22) we have $Y_t + U_i^t < \Theta(\tau)(1 - U_i^{body})$. By Lemma 7.6 we know τ_i^t is not a bottleneck.

- $U_i^{body} \geq \frac{\Theta(\tau)}{1+\Theta(\tau)}$

Since τ_i is a heavy task but not pre-assigned, it failed the pre-assign condition, satisfying the negation of that condition:

$$\sum_{j>i} U_j > (|\mathscr{P}^{\triangleright}(\tau_i)| - 1) \cdot \Lambda(\tau) \tag{7.23}$$

We split the utilization sum of all lower-priority tasks into two parts: $\mathscr{U}^{\beta}$, the part contributed by tasks on $\mathscr{P}_{\geq m}$, $\mathscr{U}^{\alpha}$, the part contributed by pre-assigned tasks on $\mathscr{P} \setminus \mathscr{P}_{\geq m}$. By the partitioning algorithm construction, we know the $\mathscr{U}^{\alpha}$ part is on processors in $\mathscr{P}^{\triangleright}(\tau_i) \setminus \mathscr{P}_{\geq m}$. We further know that each pre-assigned processor has one pre-assigned task, and each task has a utilization of at most $\Lambda(\tau)$ (our assumption stated in the beginning of Sect. 7.6). Thus, we have

$$\mathscr{U}^{\beta} \leq (|\mathscr{P}^{\triangleright}(\tau_i)| - |\mathscr{P}_{\geq m}|) \cdot \Lambda(\tau) \tag{7.24}$$

By replacing $\sum_{j>i} U_j$ by $\mathscr{U}^{\alpha} + \mathscr{U}^{\beta}$ in (7.11) and applying (7.12), we get

$$\mathscr{U}^{\alpha} > (|\mathscr{P}_{\geq m}| - 1) \cdot \Lambda(\tau) \tag{7.25}$$

The assigned utilizations on processors in $\mathscr{P}_{\geq m}$ consists of three parts: (i) the utilization of tasks with lower priority than τ_i, (ii) the utilization of τ_i, and (iii) the utilization of tasks with higher priority than τ_i. We know that part (i) is $\mathscr{U}^{\alpha}$, part (ii) is U_i, and the part (iii) is at least Y_t. So we have

$$\mathscr{U}^{\alpha} + U_i + Y_t \leq \sum_{P_q \in \mathscr{P}_{\geq m}} \mathscr{U}(P_q) \tag{7.26}$$

By (7.19), (7.25), and (7.26) we have

$$Y_t + U_i < \Lambda(\tau)$$

$$\Leftrightarrow Y_t + U_i^t < \Lambda(\tau) - U_i^{body}$$

$$\Rightarrow Y_t + U_i^t < \frac{2\Theta(\tau)}{1 + \Theta(\tau)} - U_i^{body} \quad \left(\Lambda(\tau) \leq \frac{2\Theta(\tau)}{1 + \Theta(\tau)} \right)$$

By the precondition of this case $U_i^{body} \geq \frac{\Theta(\tau)}{1 + \Theta(\tau)}$, we have

$$\frac{2\Theta(\tau)}{1 + \Theta(\tau)} - U_i^{body} \leq \Theta(\tau)(1 - U_i^{body})$$

Applying this to above we get $Y_t + U_i^t < \Theta(\tau)(1 - U_i^{body})$. By Lemma 7.6 we know τ_i^t is not a bottleneck.

In summary, we have shown that in both cases the tail subtask τ_i^t is not a bottleneck of P_m. So we can conclude that there is no bottleneck on P_m, which results in a contradiction and establishes the proof.

7.6.2.3 Utilization Bound

Lemma 7.9 (base case) and Lemma 7.11 (inductive step) inductively proved that after a *failed* partitioning, the total utilization of all processors is *at least* $M \cdot \Lambda(\tau)$. And since there are (sub)tasks not assigned to any processor after a failed partitioning, τ's normalized utilization $\mathcal{U}_M(\tau)$ is *strictly larger* than $\Lambda(\tau)$. So we can conclude:

Lemma 7.12. *Given a task set τ and a* D-PUB $\Lambda(\tau) \leq \frac{2\Theta(\tau)}{1+\Theta(\tau)}$. τ *can be successfully partitioned by* SPA2 *if its normalized utilization $\mathcal{U}_M(\tau)$ is bounded by $\Lambda(\tau)$.*

Now we will show that a task set is guaranteed to be schedulable if it is successfully partitioned by SPA2.

Lemma 7.13. *If a task set is successfully partitioned by* SPA2, *the tasks on each processor are schedulable by* RMS.

Proof. SPA2 uses routine Assign for task assignment and splitting, which assumes a body subtask has the highest priority on its host processor (this has been shown to be true for SPA1 in Lemma 7.2, so in SPA1 a successfully partitioning implies the schedulability). In SPA2, this assumption is clearly true for *normal* processors, on which the task assignment is exactly the same as SPA1. In the following, we will show this assumption is also true for *pre-assigned* processors.

Let P_q be a pre-assigned processor involved in the third phase of SPA2, and a body subtask $\tau_i^{b_j}$ is assigned to P_q. By Lemmas 7.9 and 7.11 we can inductively prove that the total utilization of processors in $\mathscr{P}_{\geq q+1}$ is at least $|\mathscr{P}_{\geq q+1}| \cdot \Lambda(\tau)$. So by Lemma 7.10 we know a pre-assigned task on processors P_q has the lowest priority on that processor, particularly, has lower priority than $\tau_i^{b_j}$. We also know that all other tasks on P_q have lower priority than $\tau_i^{b_j}$, since tasks are assigned in increasing priority order and $\tau_i^{b_j}$ is the last one assigned to P_q.

In summary we know that after partitioned by SPA2, any body subtask has the highest priority on its host processor. So Assign indeed performs a correct task assignment and splitting, which guarantees that all deadlines can be met at run-time.

By now we have proved that any task with total utilization no larger than $\Lambda(\tau)$ can be successfully partitioned by SPA2, and all tasks can meet deadline if they are scheduled on each processor by RMS. So we can conclude the utilization bound of SPA2:

Theorem 7.2. *Given a parametric utilization bound* $\Lambda(\tau) \leq \frac{2\Theta(\tau)}{1+\Theta(\tau)}$ *derived from the task set* τ*'s parameters. If*

$$\mathscr{U}_M(\tau) \leq \Lambda(\tau)$$

then τ *is schedulable by* **SPA2**.

Proof. Directly follows Lemmas 7.12 and 7.13.

7.7 Conclusions

We have developed new fixed-priority multiprocessor scheduling algorithms overstepping the Liu and Layland utilization bound. The first algorithm **SPA1** can achieve any sustainable parametric utilization bound for light task sets. The second algorithm **SPA2** gets rid of the light restriction and work for any task set, if the bound is under a threshold $\frac{2\Theta(\tau)}{1+\Theta(\tau)}$. Further, the new algorithms use exact analysis RTA, instead of the worst-case utilization threshold as in [68], to determine the maximal workload assigned to each processor. Therefore, the average-case performance is significantly improved. As future work, we will extend our algorithms to deal with task graphs specifying task dependencies and communications.

Chapter 8
Cache-Aware Scheduling

The major obstacle to use multicores for real-time applications is that we may not predict and provide any guarantee on real-time properties of embedded software on such platforms; the way of handling the on-chip shared resources such as L2 cache may have a significant impact on the timing predictability. In this chapter, we propose to use cache space isolation techniques to avoid cache contention for hard real-time tasks running on multicores with shared caches. We present a scheduling strategy for real-time tasks with both timing and cache space constraints, which allows each task to use a fixed number of cache partitions, and makes sure that at any time a cache partition is occupied by at most one running task. In this way, the cache spaces of tasks are isolated at run-time.

As technical contributions, we present solutions for the scheduling analysis problem. For simplicity, the presentation will focus on non-preemptive fixed-priority scheduling. However our techniques can be easily adapted to deal with other scheduling strategies like EDF. We have developed a sufficient schedulability test for non-preemptive fixed-priority scheduling for multicores with shared L2 cache, encoded as a linear programming problem. To improve the scalability of the test, we then develop our second schedulability test of quadratic complexity, which is an over approximation of the first test. To evaluate the performance and scalability of our techniques, we use randomly generated task sets. Our experiments show that the first test which employs an LP solver can easily handle task sets with thousands of tasks in minutes using a desktop computer. It is also shown that the second test is comparable with the first one in terms of precision, but scales much better due to its low complexity, and is therefore a good candidate for efficient schedulability tests in the design loop for embedded systems or as an on-line test for admission control.

© Springer International Publishing Switzerland 2016
N. Guan, *Techniques for Building Timing-Predictable Embedded Systems*,
DOI 10.1007/978-3-319-27198-9_8

8.1 Introduction

It is predicted that multicores will be increasingly used in future embedded systems for high performance and low energy consumption. The major obstacle is that we may not predict and provide any guarantee on real-time properties of embedded software on such platforms due to the on-chip shared resources. Shared caches such as L2 cache are among the most critical resources on multicores, which severely degrade the timing predictability of multi-core systems due to the cache contention between cores.

For single processor systems, there are well-developed techniques [4] for timing analysis of embedded software. Using these techniques, the worst-case execution time (WCET) of real-time tasks may be estimated, and then used for system-level timing analyses like schedulability analysis. One major problem in WCET analysis is how to predict the cache behavior, since different cache behaviors (cache hit or miss) will result in different execution times of each instruction. The cache behavior modeling and analysis for single-processor architectures have been intensively studied in the past decades and are supported now in most existing WCET analysis tools [4]. Unfortunately the existing techniques for single processor platforms are not applicable for multicores with shared caches. The reason is that a task running on one core may evict the useful L2 cache content belonging to a task running on another core and therefore the WCET of one task cannot be estimated in isolation from the other tasks as for single processor systems. Essentially, the challenge is to model and predict the cache behavior for concurrent programs (not sequential programs as for the case of single processor systems) running on different cores.

To our best knowledge, the only known work on WCET analysis for multicores with shared cache is [89], which is only applicable to a very special application scenario and very simple hardware architecture (we will discuss its limitation in Sect. 8.2). Researchers in the WCET analysis community agree that "it will be extremely difficult, if not impossible, to develop analysis methods that can accurately capture the contention among multiple cores in a shared cache" [2].

The goal of this chapter is not to solve the above challenging problem. Instead, we use cache partitioning techniques such as page-coloring [165] combined with scheduling to isolate the cache spaces of hard real-time tasks running simultaneously to avoid the interference between them. This yields an efficient method— cache space isolation—to control the shared cache access, in which a portion of the shared cache is assigned to each running task, and the cache replacement is restricted to each individual partition. For single-processor multi-tasking systems, cache space isolation allows compositional timing analysis where the WCET of tasks can be estimated separately using existing WCET analysis techniques [166]. For multicores, to enable compositional timing analysis, we need isolation techniques for all the shared resources. For the on-chip bus bandwidth, techniques such as time-slicing have been studied in, e.g., [167]. In this chapter, we shall focus on shared caches only, and study the scheduling and analysis problem for hard real-time tasks with timing and cache space constraints, on multicores with shared L2 cache.

We assume that the shared cache is divided into partitions, and further assume that the cache space size of each application task has been estimated by, for example, the miss-rate/cache-size curve or static analysis, and the WCET of each task is obtained with its assigned cache space size. In the system design phase, one can adjust tasks' L2 cache space sizes (and therefore their WCETs) to improve the system real-time performance, which can be built upon the schedulability analysis techniques studied in this work.

We shall present a cache-aware scheduling algorithm which makes sure that at any time, any two running tasks' cache spaces are non-overlapped. A task can get to execute only if it gets an idle core as well as enough space (not necessarily continuous) on the shared cache. For the simplicity of presentation, we shall focus on non-preemptive fixed-priority scheduling. However, our results can be easily adapted to other scheduling strategies such as EDF. Our first technical contribution is a sufficient schedulability test for multicores with shared L2 cache, encoded as a linear programming problem. To improve its scalability, we then propose our second schedulability test of quadratic complexity, which is an over approximation of the first test. To evaluate the performance and scalability of our techniques, we use randomly generated task sets. Our experiments show that the first test which employs an LP solver can easily handle task sets with thousands of tasks in minutes using a desktop computer. It is also shown that the second test is comparable with the first one in terms of precision, but scales much better due to its low complexity, and therefore it is a good candidate for efficient schedulability tests in the design loop for embedded systems.

8.2   Related Work

Since L2 misses affect the system performance to a much greater extent than L1 misses or pipeline conflicts [168], the shared cache contention may dramatically degrade the system performance and predictability. Chandra et al. [169] showed that a thread's execution time may be up to 65 % longer when it runs with a high-miss-rate co-runner than with a low-miss-rate co-runner. Such dramatic slowdowns were due to significant increases in L2 cache miss rates experienced with a high-miss-rate co-runner, as opposed to a low-miss-rate co-runner.

L2 contention can be reduced by discouraging threads with heavy memory-to-L2 traffic from being co-scheduled [168]. Anderson et al. [170–172] applied the policy of encouraging or discouraging the co-scheduling of tasks (or jobs), to improve the cache performance and also to meet the real-time constraints. All these works assumed that the WCETs of real-time threads are known in advance. However, although improved cache performance can directly reduce average execution costs, it is still unknown how to obtain the WCET of each real-time thread in their system model. Yan and Zhang [89] is the only known work to study the WCET analysis problem for multi-core systems with shared L2 cache. A particular scenario is assumed that two tasks simultaneously run on a dual-core processor with a

direct-mapped shared L2 instruction cache. However, their analysis technique is
quite limited: firstly, most of today's multi-core processors employ set-associative
caches rather than direct-mapped cache as their L2 cache; secondly, when the system
contains more cores and more tasks, their analysis will be extremely pessimistic;
thirdly, their analysis technique cannot handle tasks in priority-driven scheduling
systems.

In contrast with Anderson's work, we employ cache space isolation in the
scheduling algorithms to avoid the cache accessing interference between tasks
simultaneously running on different cores, and therefore we can apply existing
analysis techniques to derive safe upper bounds of a task's WCET,[1] with which
we can do safe schedulability analysis for the task system.

The schedulability analysis problem of global multiprocessor scheduling has
been intensively studied [50, 52–54, 134, 138]. These analysis techniques are also
extended to deal with more general cases, e.g., the global scheduling on 1-D FPGAs
[173, 174], where a task may occupy multiple resources (columns on FPGAs) during
execution. However, all these techniques are not applicable to our problem, since
with cache space isolation, tasks are actually scheduled on two resources: cores and
the shared cache.

Fisher et al. [175] studied the problem of static allocation of periodic tasks onto
a multiprocessor platform such that on each processor, the total utilization of the
allocated tasks is no larger than 1, as well as the total memory size of the allocated
tasks does not exceed the processor's memory capacity. Suhendra et al. [176] and
Salamy et al. [177] studied the problem of how to statically allocate and schedule
a task graph onto an MPSoC, in which each processor has a private scratch-pad
memory, to maximize the system throughput. In summary, in the above work tasks
are statically allocated to processors, so the schedulability analysis problem is trivial
(reduced to the case of single-processor scheduling). In our work, different instances
of a task are allowed to run on different cores, so the schedulability analysis
problem is more difficult. In [2], several scheduling policies with shared cache
partitioning and locking are experimentally evaluated, however, the schedulability
analysis problem was not studied.

8.3   Preliminaries

In this section, we briefly describe the basic assumptions on the hardware platform
and application tasks, which our work is based on.

---

[1] We focus on the interference caused by the shared L2 cache, and there could be other interference
between tasks running simultaneously. However, we believe the scheduling algorithm and analysis
techniques in this paper is a necessary step towards completely avoiding interference between
tasks running on multicores, and can be integrated with techniques of performance isolation on
other shared resources, for instance, the work in [167] to avoid interference caused by the shared
on-chip bus.

8.3.1 Cache Space Partitioning

We assume a multi-core containing a fixed number of processor cores sharing an on-chip cache. Note that this is usually an L2 cache. We will not explicitly model core-local caches (usually L1) or other shared resources like interconnects. Since concurrent accesses to the shared cache give raise to the problem of reduced predictability due to cache interference, we assume the existence of a cache partitioning mechanism, allowing to divide the cache space into non-overlapping partitions for independent use by the computation tasks, see Fig. 8.1a.

Partitioning a cache shared among several tasks at the same time is a concept which has already been used, most notably, for reducing interference in order to improve average-case performance or to increase predictability in single-core settings with preemption [178–180].

Different approaches may be used to achieve cache partitioning. Assuming a k-associative cache that consists of l cache sets with k cache lines each, one can distinguish *set-based* [165] and *associativity-based* [181] partitioning. The first one is also called *row-based partitioning* and assigns different cache sets to different partitions. It therefore enables up to l partitions and is thus quite fine-grained for bigger caches. The second one assigns a certain amount of lines within each cache set to different partitions and is also called *column-based partitioning*, so it is rather coarse-grained with a maximum of just k partitions. Mixtures of both variants are also possible. The approaches can be software- or hardware-based and differ

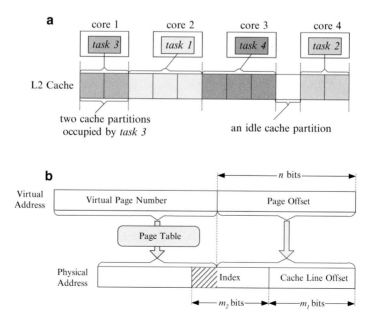

Fig. 8.1 Cache space isolation and page coloring. (**a**) Cache space isolation. (**b**) Address mapping

regarding additional hardware requirements, partitioning granularity, influence on memory layout and the possibility as well as complexity of on-line repartitioning.

Here we give a brief description of a set-based approach, which is also known as *page coloring*. It has the advantage of being entirely software-based by exploiting the translation from virtual to physical memory addresses present in the virtual memory system.[2] Assume a simple hardware-indexed cache with cache line size of 2^{m_1} words and 2^{m_2} cache sets, so the least significant bits of the physical address will contain m_1 bits used as cache line offset and m_2 bits used as the set number, see Fig. 8.1b. Further assume a virtual page size of 2^n words, so the n least significant bits of the virtual address comprise the page offset. Consequently, all the other (more significant) bits are the page number and will be translated by the virtual memory system via the page table into the most significant bits of the physical address. If $m_1 + m_2 > n$ (which is the case with larger caches), a certain number of bits used to address the cache set are actually "controlled" by the virtual memory system, so that each virtual page can be (indirectly) mapped on a particular subset of all cache sets. The number of available *page colors* by that method is therefore $2^{(m_1+m_2)-n}$.

An example system supporting cache partitioning is reported in [182], where the authors modified the Linux kernel to support page-coloring based cache space isolation, in which 16 colors are supported, and conducted intensive experiments on a Power 5 dual-core processor. Note that the method enforces a certain (physical) memory layout, since it influences the choice of physical addresses. This restricts the memory size available to each task, as well as flexibility for recoloring. These problems can be compensated for by a simple rewiring trick as described in [183]. Therefore it is reasonable for our model to assume a cache with equally sized cache partitions that can be assigned and reassigned arbitrarily during the lifetimes of the tasks in question.

8.3.2   Task Model

Assume a multi-core platform consisting of M cores and A cache partitions, and a set τ of independent sporadic tasks whose numbers of cache partitions (cache space size needed) and WCETs are known for the platform. We use $\tau_i = \langle A_i, C_i, D_i, T_i \rangle$ to denote such a task where A_i is the *cache space size*, C_i is the *WCET*, $D_i \leq T_i$ is the relative deadline for each release, and T_i is the minimum inter-arrival separation time also referred to as the *period* of the task. We further assume that all tasks are ordered by priorities, i.e., τ_i has higher priority than τ_j iff $i < j$. The *utilization* of a task τ_i is $U_i = C_i/T_i$ and its *slack* $S_i = D_i - C_i$, which is the longest delay allowed before actually running without missing its deadline.

---

[2]Note that this is just an example of how cache partitioning can be achieved; by no means is virtual memory a necessity to the results presented in this chapter.

A sporadic task τ_i generates a potentially infinite sequence of *jobs* with successive job-arrivals separated by at least T_i time units. The αth job of task τ_i is denoted by J_i^α, so we can denote a job sequence of task τ_i with $(J_i^1, J_i^2, \ldots)$. We omit α and just use J_i to denote a job of τ_i if there is no need to identify which job it is. Each job J_i adheres to the conditions A_i, C_i and D_i of its task τ_i and has additional properties concerning *absolute time points* related to its execution, which we denote with lowercase letters: The *release time*, denoted by r_i, the *deadline*, denoted by d_i and derived using $d_i = r_i + D_i$, and the *latest start time*, denoted by l_i and derived using $l_i = r_i + S_i$. Finally, without losing generality, we assume that time is dense in our model.

8.4 Cache-Aware Scheduling

We present the basic scheduling algorithm studied in this chapter, and the analysis framework for the technical contributions presented in the next sections. We should point out that the simple scheduling algorithm itself is not the main contribution of this work. Our contributions are in solving the schedulability problem for this algorithm.

8.4.1 The Scheduling Algorithm FP$_{CA}$

Since cache-related context-switch overhead of each task due to preemption is usually hard to predict, we focus on non-preemptive scheduling. The idea of cache space isolation can be applied to many different traditional multiprocessor scheduling algorithms, and for simplicity reasons, we will take the Non-preemptive Fixed-Priority Scheduling as the example in this chapter.

The algorithm, the Cache-Aware Non-preemptive Fixed-Priority Scheduling (FP$_{CA}$), is executed whenever a job finishes or when a new job arrives. It always schedules the highest priority waiting job for execution, if there are enough resources available. In particular, a job J_i is scheduled for execution if:

1. J_i is the job of highest priority among all waiting jobs,
2. There is at least one core idle, and
3. Enough cache partitions, i.e. at least A_i, are idle.

Note that, since we suppose $D_i \le T_i$ for each task, there is at most one job of each task at any time instant.

Figure 8.2 shows an example of the task set in Table 8.1 scheduled by FP$_{CA}$ (the scenario that all tasks are released together). Note that at time 0, the job J_4^1 cannot execute according to the definition of FP$_{CA}$, although it is ready and there is an idle core and enough idle cache partitions to fit it, since it is not at the first position of the waiting queue, i.e. there is a higher priority job (J_3^1) waiting for

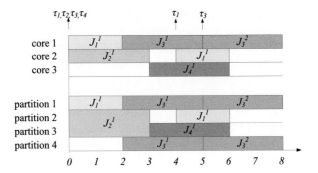

Fig. 8.2 An example to illustrate FP$_{CA}$

Table 8.1 The task set to illustrate FP$_{CA}$

Task	D_i	T_i	C_i	A_i
τ_1	3	3	2	1
τ_2	4	4	3	2
τ_3	5	5	2	2
τ_4	8	8	2	1

execution. J_3^1 cannot execute since there is not enough idle cache partitions available. Thus, we note that FP$_{CA}$ may *waste resources* as it does not schedule lower priority ready jobs to execute in advance of higher priority ready jobs even though there are enough resources available to accommodate them. However, it enforces a stricter priority ordering, which is in general good for predictability. We name this kind of scheduling policy as *blocking-style scheduling*.

Sometimes one may prefer to allow lower priority ready jobs to execute in advance of higher priority ready jobs, if the idle cache partitions are not enough to fit the higher priority ones, to trade predictability for better resource utilization. We name this kind of scheduling policy as *non-blocking-style scheduling*.

For simplicity, we will present the schedulability analysis in context of FP$_{CA}$, which is *blocking-style scheduling*. However, note that the schedulability analysis techniques are applicable to both *blocking-style scheduling* and *non-blocking-style scheduling*. Later in Sect. 8.8, we will discuss the comparison between them in more detail.

8.4.2 Problem Window Analysis

To check whether a given set of tasks can be scheduled using the above algorithm without missing the deadline for any job released, we shall study the time interval during which an assumed deadline missing task is prevented from running. Note that this interval is the so-called slack of the task, which we shall also call the *problem window* [52].

In the following, we outline how the problem window can be used for schedulability analysis in the case when tasks are scheduled only on the cores or only on the shared cache partitions. Two schedulability test conditions will be developed for the two special cases. Then, in Sect. 8.5, we combine them to deal with the general case.

The Case Without Cache Scheduling

A schedulability test for the case when the tasks are scheduled only on the cores can be derived as follows:

1. Assume M cores for execution of a task set τ as described before, but in the task model, the A_i's are 0 (alternatively the total number of cache partitions is large enough such that no task will be blocked by a busy cache).
2. Suppose that the task set τ is unschedulable, then there is a job sequence $(J_{k_1}^{\alpha_1}, J_{k_2}^{\alpha_2}, \ldots)$ in which a job misses its deadline. Let J_k, a job of τ_k, be the first job missing its deadline. Its release time is r_k and the latest time point, at which it would have needed to start running (but it did not, since it is missing its deadline) is $l_k = r_k + S_k$. We define the time interval $[r_k, l_k]$ of length S_k as the *problem window*, as shown in Fig. 8.3a. The intuition is that at all time points within the interval, each of the cores must be occupied by another task, preventing J_k from running.
3. To find out why J_k is not scheduled to run during the window, we may estimate the workload or an upper bound of this, generated by a task that may occupy a core in the window. We denote such an upper bound by I_k^i, which is normally called the interference contributed to J_k's problem window by task τ_i. The sum $\sum_i I_k^i$ is an upper bound of the total workload interfering with J_k in the problem window. We describe in detail how to calculate such an upper bound in the next section. A more precise calculation is given in the appendix.
4. We note that the non-preemptive fixed-priority scheduling algorithm (without cache) enjoys the *work-conserving* property, that is, none of the M cores is idle if there is some ready job waiting for execution. Therefore, J_k can miss its deadline only if $\sum_i I_k^i \geq S_k \cdot M$ holds, i.e., the whole area with diagonals in Fig. 8.3a is occupied. Otherwise, J_k is safe from ever missing its deadline, i.e., τ_k is schedulable, if the following condition holds: $\sum_i I_k^i < S_k \cdot M$.

We may also view this last step in a different way. We know that the sum of all work (of all tasks τ_i) interfering with J_k is bounded by $\sum_i I_k^i$, and it is in the worst case executed in parallel on M cores, thus preventing J_k from running. Therefore, if we divide this sum $\sum_i I_k^i$ by M, we get an upper bound on the maximum time that job J_k can be delayed by other tasks. We call this the *interference time*. Consequently, J_k is guaranteed to be schedulable, if this interference time is strictly less than its slack, i.e., if the following condition holds:

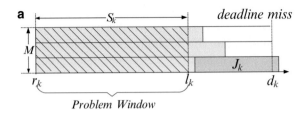

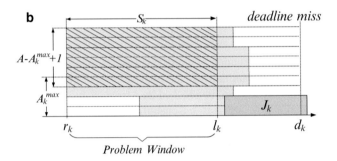

Fig. 8.3 Problem window. (**a**) Only considering scheduling on cores. (**b**) Only considering scheduling on the cache

$$\frac{1}{M}\sum_i I_k^i < S_k \qquad (8.1)$$

By applying the above procedure to each task $\tau_k \in \tau$ (i.e., checking that the inequality holds for all tasks), one can construct a sufficient schedulability test for the case without a shared cache.

The Case Without Core Scheduling

The above problem window analysis can be generalized to the case where each task occupies *several* computing resources. In our scenario, one task can occupy *several* cache partitions at once while executing.

To present the idea, let us assume for the moment that we *only* care about the scheduling of the shared cache (suppose there are always enough cores for tasks to execute). A job J_i may start running as soon as it is the first one in the waiting queue Q_{wait}, and the number of idle cache partitions is at least A_i. Otherwise, J_k in Q_{wait} (it is now not necessarily the first job) may have to wait, if the number of idle cache partitions is less than $\max(A_1, \dots, A_k)$.

Note that we take the maximum over all higher priority tasks here, since even though there might be A_k cache partitions idle, there could be a job J_i of higher priority ($i < k$) in Q_{wait} that needs more cache partitions to run, but is prevented from running, which in turn prevents J_k from running, because of the blocking property of FP_{CA}. We define this number as

$$A_k^{\max} = \max_{i \leq k} A_i.$$

Note that $A_k^{\max}$ is the minimal number of idle cache partitions needed in order that J_k is not blocked from running because of a busy cache. Equivalently, the minimal number of busy cache partitions that may block J_k from running is $A - A_k^{\max} + 1$.

Therefore J_k can miss its deadline only if the whole area with diagonals in Fig. 8.3b is occupied. Since each task τ_i is occupying A_i cache partitions while it is executing, we know that J_k can miss its deadline only if the condition $\sum_i A_i I_k^i \geq S_k \cdot (A - A_k^{\max} + 1)$ holds. Thus we have a test condition for scheduling analysis when the shared cache is considered: $\sum_i A_i I_k^i < S_k \cdot (A - A_k^{\max} + 1)$.

Like in Sect. 8.4.2, we again prefer the view on that in terms of *interference time*: We get an upper bound of the interference time suffered by job J_k in the problem window by dividing this sum of maximal total cache use $\sum_i A_i I_k^i$ by the minimal number of busy cache partitions $(A - A_k^{\max} + 1)$ throughout the problem window. This is, again, an upper bound of the time, by which job J_k can be delayed by other tasks. Thus, the schedulability test condition is

$$\frac{1}{A - A_k^{\max} + 1} \sum_i A_i I_k^i < S_k \tag{8.2}$$

As we see now in Constraints (8.1) and (8.2), one can derive test conditions for scheduling on cores and cache partitions *separately*, once I_k^i is known for each task τ_i. Since in the scheduling algorithm in question, FP_{CA}, the scheduling happens on cores and cache *together*, the conditions have to be combined in a way that still makes for a safe schedulability test condition. In the following section, we will derive a novel way of combining both conditions.

8.5   The First Test: LP-Based

In order to apply the problem window analysis to FP_{CA}, two questions need to be answered:

1. *How to compute I_k^i, i.e., an upper bound of the interference of each task τ_i in the problem window?* We will answer this question in Sect. 8.5.1.
2. *How to determine whether the interference of all tasks is large enough to prevent J_k from executing in the problem window?* We will answer this question in Sect. 8.5.2.

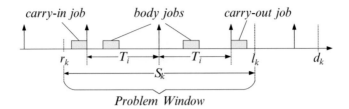

Fig. 8.4 Carry-in job, carry-out job, and body jobs

8.5.1   Interference Calculation

The first question can be answered by categorizing each job of τ_i in the problem window into one of three types, as shown in Fig. 8.4:

body job: a job with both release time and deadline *in* the problem window; All the body jobs together contribute $\lfloor S_k/T_i \rfloor \cdot C_i$ to the interference.

carry-in job: a job with release time *earlier than* r_k, but with deadline *in* the problem window; This job contributes at most C_i to the interference.

carry-out job: a job with release time *in* the problem window, but with deadline *later* than l_k; This job also contributes at most C_i to the interference.

It follows that an upper bound of τ_i's interference in the problem window is given by

$$I_k^i = \left(\left\lfloor \frac{S_k}{T_i} \right\rfloor + 2 \right) \cdot C_i. \tag{8.3}$$

We can derive a more precise computation of I_k^i by carefully identifying the worst-case scenario of each task's interference, which is given in the appendix.

8.5.2   Schedulability Test as an LP Problem

The answer of the second question is non-trivial.

As introduced in Sect. 8.4.2, the problem window analysis can be applied to analyzing the scheduling on cores or on the cache *separately*. However, if we consider the scheduling on both cores and cache, it is generally unknown what the lower bound of the occupied resources on each of them is, to cause J_k to miss deadline. For example, in Fig. 8.2, at time instant 0, the job J_3^1 is ready for execution, but it cannot execute since the number of idle cache partitions on the shared cache is not enough to accommodate it, so it is not true any longer that all M cores must be busy during the problem window to cause the considered task to miss its deadline.

Dividing Up the Problem Window

The key observation to our analysis is that at each time point in the problem window where the job J_k cannot start running, all cores are already occupied, or—wherever that is not the case—enough cache partitions are occupied to prevent J_k from running. This is expressed in the following lemma about FP_{CA}:

Lemma 8.1. *Let J_k be a job that misses its deadline. Then at any time instant in the problem window $[r_k, l_k]$, at least one of the following two conditions is true:*

1. All M cores are occupied;
2. At least $A - A_k^{max} + 1$ cache partitions are occupied.

Proof. Suppose there is a time instant $t \in [r_k, l_k]$, such that both of the above two conditions do not hold. Since J_k misses its deadline, it cannot start executing in the problem window, thus also not at t. Therefore, the waiting queue Q_{wait} is not empty at t, since at least J_k is in Q_{wait}.

Let now J_i be the first job in Q_{wait} at time t. Since, for FP_{CA}, the waiting queue is ordered in strict priority order, J_i is the highest priority job waiting. By assumption, there are less than $A - A_k^{max} + 1$ cache partitions occupied, so there are at least A_k^{max} partitions available. Further, $A_i \leq A_k^{max}$ by definition, and there is an idle core by assumption. Thus, J_i must be able to execute, contradicting the assumption that it is waiting. ∎

Following these two conditions, we can now divide the problem window into two parts (see Fig. 8.5):

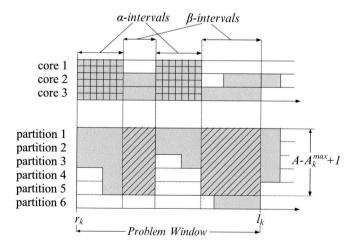

Fig. 8.5 Illustration of α-intervals and β-intervals

1. α-*intervals*, in which all cores are busy;
2. β-*intervals*, in which at least one core is idle. Note that it follows from Lemma 8.1 that during the β-intervals, at least $A - A_k^{\max} + 1$ partitions of the shared cache are occupied by a running task.

It is generally unknown at what length of the α- and β-intervals the maximal interference to J_k is achieved. We approach this by introducing a Linear Programming (LP) formulation of our problem, to create a schedulability test for FP$_{CA}$.

LP Formulation

Suppose, as before, the task set τ is unschedulable by FP$_{CA}$, and J_k is the first task that is missing its deadline. The time interval $[r_k, l_k]$ is the problem window.
The LP formulation will use the following constants:

- M: the number of cores.
- A: the total number of partitions on the shared cache.
- A_i: the number of cache partitions occupied by each task τ_i. (We also use the constant $A_k^{\max}$, which is derived from these as above.)
- I_k^i: an upper bound of the interference by τ_i in the problem window, which is computed as in Sect. 8.5.1 (or as in the appendix) for each τ_i.

Further, the following non-negative variables are used:

- α_i: for each task τ_i, we define α_i as τ_i's accumulated execution time during α-intervals.
- β_i: for each task τ_i, we define β_i as τ_i's accumulated execution time during β-intervals.

During the α-intervals, all M cores are occupied. Further, we know that $\sum_i \alpha_i$ equals to the total computation work of all tasks during the α-intervals (which is the area with grids in Fig. 8.5). We can therefore express the accumulated length of all α-intervals as

$$\frac{1}{M} \sum_i \alpha_i \tag{8.4}$$

During the β-intervals, at least $A-A_k^{\max}+1$ cache partitions are occupied. Further, $\sum_i A_i\beta_i$ is the total cache partition use of all tasks during the β-intervals (which is an upper bound of the area with diagonals in Fig. 8.5). We can therefore express an upper bound of the accumulated length of all β-intervals as

$$\frac{1}{A - A_k^{\max} + 1} \sum_i A_i\beta_i \tag{8.5}$$

Since J_k is not schedulable, we further know that the sum of the accumulated lengths of the α- and β-intervals is at least S_k. Thus, using the expressions from (8.4) and (8.5), it must hold that

$$\sum_i \left(\frac{1}{M} \alpha_i + \frac{A_i}{A - A_k^{\max} + 1} \beta_i \right) \geq S_k \tag{8.6}$$

We can use an LP solver to detect if the α_i and β_i variables can be chosen in a way to satisfy this condition. If this is not the case, then τ_k would be schedulable. We can use the object function of our LP formulation for that check:

$$Maximize \quad \sum_i \left(\frac{1}{M} \alpha_i + \frac{A_i}{A - A_k^{\max} + 1} \beta_i \right) \tag{8.7}$$

Thus, if the solution of the LP problem is smaller than S_k, we can determine that τ_k is schedulable.

So far, the variables α_i and β_i are not bounded, so without further constraints, the LP formulation will not have a bounded solution (which would trivially render all tasks unschedulable). Therefore, we add constraints on the free variables that follow directly from the structure of our schedulability problem. We have three constraints:

φ_1: Interference Constraint   We know that I_k^j is the upper bound of the work done by τ_j in the problem window, so we have

$$\forall j : \alpha_j + \beta_j \leq I_k^j$$

φ_2: Core Constraint   The work done by a task in the α-intervals cannot be larger than the total accumulated length of the α-intervals [see Expression (8.4)], so we have

$$\forall j : \alpha_j \leq \frac{1}{M} \sum_i \alpha_i$$

φ_3: Cache Constraint   The work done by a task in the β-intervals cannot be larger than the total accumulated length of the β-intervals. Thus, it cannot be larger than the upper bound of the total length of the β-intervals [see Expression (8.5)], so we have

$$\forall j : \beta_j \leq \frac{1}{A - A_k^{\max} + 1} \sum_i A_i \beta_i$$

To test τ_k for schedulability, we can now invoke an LP solver on the LP problem defined by constraints φ_1 to φ_3 and the object function in (8.7). By construction, we have a first schedulability test for τ:

Theorem 8.1 (The First Test). *For each task τ_k, let χ_k denote the solution of the LP problem shown above. A task set τ is schedulable by FP$_{CA}$, if for each task $\tau_k \in \tau$ it holds that*

$$\chi_k < S_k. \tag{8.8}$$

8.6 The Second Test: Closed Form

Although the LP-based test presented in the previous section exhibits quite good scalability (as will be shown in Sect. 8.7), simple test conditions are often preferred in, e.g., on-line admission control and efficient analysis in the systems design loop. Thus, we will present a second schedulability test, which can be seen as an over-approximation of the LP-based test. It has quadratic computational complexity.

In the LP-based test, each task τ_i finds its interference I_k^i divided into two parts α_i and β_i, expressed by constraint φ_1. From the object function (8.7) one can see that if $1/M \geq A_i/(A - A_k^{\max} + 1)$, τ_i tends to contribute with as much α_i as possible, as long as φ_2 is respected; likewise in the opposite case with β_i. However, since the accumulated lengths of the α- and β-intervals (as used on the right-hand sides of φ_2 and φ_3) are also dependent on the unknown variables, it is in general unknown how the α_i and β_i variables are chosen to maximize the interference time caused by all tasks. Therefore, the first schedulability test employs the LP solver to help "searching over all possible cases" for the maximal solution.

However, if we take out the constraints φ_2 and φ_3 from the LP formulation, each task τ_i will contribute $\alpha_i = I_k^i, \beta_i = 0$ if $1/M \geq A_i/(A - A_k^{\max} + 1)$, or $\beta_i = I_k^i, \alpha_i = 0$ otherwise, to maximize the object function. With this observation, we can derive a closed-form schedulability test which does not need to solve a search problem:

Theorem 8.2 (The Second Test). *For each task τ_k let*

$$\chi_k^* := \sum_i \max\left(\frac{1}{M}, \frac{A_i}{A - A_k^{\max} + 1}\right) \cdot I_k^i.$$

A task set τ is schedulable by FP$_{CA}$, if for each task $\tau_k \in \tau$ it holds that

$$\chi_k^* < S_k. \tag{8.9}$$

Proof. We prove the theorem indirectly. Let τ be a task set not schedulable by FP$_{CA}$, and J_k the deadline missing task as before. We already know that this implies the existence of a solution for the LP problem, such that in particular, φ_1 to φ_3 hold, and the value of the object function satisfies the following inequality:

$$\sum_i \left(\frac{1}{M} \cdot \alpha_i + \frac{A_i}{A - A_k^{\max} + 1} \cdot \beta_i\right) \geq S_k$$

Table 8.2 The example task
set to illustrate Theorem 8.2

M	$A - A_k^{\max} + 1$	I_k^1	A_1	I_k^2	A_2	I_k^3	A_3
2	4	4	1	4	3	6	1

By relaxing the inequality, we get

$$\sum_i \max\left(\frac{1}{M}, \frac{A_i}{A - A_k^{\max} + 1}\right) \cdot (\alpha_i + \beta_i) \geq S_k$$

Now, we apply condition φ_1:

$$\underbrace{\sum_i \max\left(\frac{1}{M}, \frac{A_i}{A - A_k^{\max} + 1}\right) \cdot I_k^i \geq S_k}_{\chi_k^*}$$

The theorem follows.

Note that the upper bound χ_k^* derived in the above theorem is an over-approximation of the LP solution χ_k in the previous section. For example, consider a task set with the interference parameters as stated in Table 8.2. The LP problem (from the first test) has the following solution:

$$\alpha_1 = 4 \ \beta_1 = 0$$
$$\alpha_2 = 1 \ \beta_2 = 3$$
$$\alpha_3 = 3 \ \beta_3 = 3$$

This results in an upper bound of $\chi_k = 7$, which is the value of the object function. In the second test, for χ_k^*, each task τ_i contributes all I_k^i as α_i if $1/M > A_i/(A - A_k^{\max} + 1)$, and as β_i otherwise, so we get the following bound:

$$\chi_k^* = \frac{1}{2} \cdot 4 + \frac{3}{4} \cdot 4 + \frac{1}{2} \cdot 6 = 8$$

Although the simple test condition is more pessimistic than the LP-based test, we found by extensive experiments that the performance of the second test is very close to the LP-based test in terms of acceptance ratio. We will show that in Sect. 8.7. It follows that, for practical matters, the second test does not lose much precision for most task sets, while being of comparatively low complexity (quadratic with respect to the number of tasks).

8.7   Performance Evaluation

At first we evaluate the performance of the proposed schedulability tests in terms of acceptance ratio. We follow the method in [135] to generate task sets: A task set of $M + 1$ tasks is generated and tested. Then we iteratively increase the number of tasks by 1 to generate a new task set, and all the schedulability tests are run on the new task set. This process is iterated until the total processor utilization exceeds M. The whole procedure is then repeated, starting with a new task set of $M + 1$ tasks, until a reasonable sample space has been generated and tested. This method of generating random task sets produces a fairly uniform distribution of total utilizations, except at the extreme end of low utilization.

Figure 8.6 shows the acceptance ratio of the first test (denoted by "T-1"), and second test (denoted by "T-2") and the simulation (denoted by "Sim"). Since it is not computationally feasible to try all possible task release offsets and inter-release separations exhaustively in simulations, all task release offsets are set to be zero and all tasks are released periodically, and simulation is run for the hyper-period of all task periods. Simulation results obtained under this assumption may sometimes determine a task set to be schedulable even though it is not, but they can serve as a coarse upper bound of the acceptance ratio.

The parameter setting in Fig. 8.6a is as follows: the number of cores is 6; the number of cache partitions is 40; for each task τ_i, T_i is uniformly distributed in $[10, 20]$, U_i is uniformly distributed in $[0.1, 0.3]$ and A_i is uniformly distributed in $[1, 5]$, and we set $Di = Ti$. We can see that the performance of the first test is a little better than the second test. In Fig. 8.6b, the range of U_i is changed to $[0.1, 0.6]$ and other settings are the same as Fig. 8.6a. In Fig. 8.6c, the range of A_i is changed to $[2, 10]$ and other settings are the same as Fig. 8.6a. We can see that, in both cases, the acceptance ratio of all the simulations and tests degrades a little bit when the average utilization of tasks is slightly decreased, and the difference between the performance of the first test and second test is even smaller. In summary we can see that the second test does not lose too much precision, compared to the first test condition.

As mentioned earlier, the second test is of $O(N^2)$ complexity. The scalability of the first test is of our special concern since it employs the LP formulation. We use

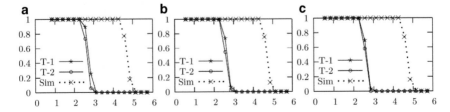

Fig. 8.6 Acceptance ratio: X-axis is total utilization $\sum_i U_i$; Y-axis is acceptance ratio. (**a**) $0.1 \leq U_i \leq 0.3, 1 \leq A_i \leq 5$. (**b**) $0.1 \leq U_i \leq 0.6, 1 \leq A_i \leq 5$. (**c**) $0.1 \leq U_i \leq 0.3, 2 \leq A_i \leq 10$

Table 8.3 Running time and peak memory usage of lpsolve to solve the LP formulation in the first test

Number of tasks	4000	6000	8000	10,000
Time in LP (s)	49.24	114.53	208.45	334.95
Mem. in LP (KB)	20,344	28,876	37,556	46,664

the open source LP solver lpsolve [125] to solve the LP formulation of the first test. Table 8.3 shows the running time and maximal peak memory usage of lpsolve with different task set scales. The experiment is conducted on a normal desktop computer with an Intel Core2 processor (2.83 GHz) and 2G memory. The experiments show that the first test can handle task sets with thousands of tasks in minutes.

8.8   Extensions

8.8.1   Blocking vs. Non-blocking Scheduling

As we mentioned in Sect. 8.4.1, FP_{CA} may introduce a type of resource wasting in certain situations, caused by a difference in tasks cache requirements, in combination with strict adherence to priority ordering. The possible scenario is that when the current idle cache is not enough to fit the first job in the waiting queue (the highest-priority job among all ready jobs), there could be some lower-priority job in the waiting queue with a fitting cache requirement. In FP_{CA} as analyzed so far, lower-priority waiting jobs are not allowed to start execution in advance of the first job in the waiting queue since the priority ordering is enforced strictly. We call this sort of policy the *blocking-style* scheduling. The blocking-style scheduling would waste computing resources to guarantee the execution order of waiting jobs, as we saw earlier in the example in Fig. 8.2. But it will not suffer from the unbounded priority inversion problem due to the sharing of cache partitions as for the non-blocking approach described below.

Alternatively, to improve resource utilization, a lower-priority waiting job may be allowed to start execution in advance of the first job in the waiting queue, if the above described situation occurs, which we call *non-blocking-style* scheduling. Figure 8.7 shows how the task set in Table 8.1 is scheduled by the non-blocking-style version of FP_{CA}. In this variant, the scheduler always runs the highest priority waiting job in the queue among all jobs that can actually run, given their resource (i.e., cache) constraints. This is done until there are no more jobs of that kind. In the example in Fig. 8.7, we can see that at time instant 0, job J_4^1 starts execution although J_3^1 cannot start.

From the predictability point of view, the blocking-style scheduling is usually to be preferred, in which waiting jobs start execution in strict priority orders. The reason is that τ_h may suffer more interference than it is the case in the blocking-style scheduling, since in the non-blocking-style scheduling, a lower priority task τ_l can

Fig. 8.7 The non-blocking
version of FP_{CA}

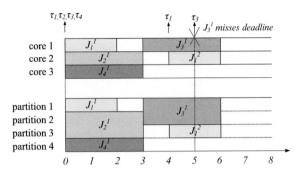

execute earlier than a higher priority task τ_h, and must run to completion because of
non-preemptive scheduling. As shown in Fig. 8.7, due to the advanced execution of
J_4^1, the start time of J_3^1 is delayed to time 3, and J_3^1 will finally miss its deadline. In the
worst case, this priority inversion effect could even cause unbounded interference to
a task with even the highest priority.

On the other hand, the non-blocking-style scheduling utilizes resources better,
since it always tries to utilize the computing resources as much as possible.
Regarding the complexity at run-time, this comes with the cost that the scheduler
needs to keep track of more than the head of the priority queue, since lower priority
tasks might be able to run. The blocking-style variant is more lightweight, since
only the head of the priority queue needs to be checked.

The system designer can choose to use blocking-style or non-blocking-style
scheduling, as well as some compromise policy as a mixture of these two alterna-
tives, according to the application requirement. However, even though the proposed
schedulability analysis techniques are done in the context of the blocking-style
scheduling, they are also applicable to the non-blocking-style scheduling. For this, it
is only necessary to incorporate some extra consideration of the interference caused
by the lower priority jobs that may execute in advance of the analyzed job. We omit
the detailed presentation of the analysis of non-blocking-style scheduling here due
to page limitations, and referred the interested readers to our technical report [184].

8.8.2   Other Scheduling Strategies

Besides FP_{CA}, cache space isolation can also be applied to other scheduling
algorithms, like EDF scheduling. The schedulability analysis in that case can be
achieved by techniques which are similar to those introduced in this chapter.

One can also apply cache space isolation in a way similar to the partitioned
multiprocessor scheduling [48]: each task is assigned to a core and a set of cache
partitions in advance. One reason for us to be interested in the partitioned scheduling
is that the shared cache on multicores could be non-uniform in terms of accessing
speed: data residing in the part of a large cache close to the core could be accessed
much faster than data residing physically farther from the core. In [185], it was

shown in an example that in a 16-megabyte on-chip L2 cache built in a 50-nm processor technology, the closest bank could be accessed in 4 cycles while an access to the farthest bank might take 47 cycles. Therefore, with non-uniform shared cache, one should assign fixed cores and cache blocks to each task for scheduling and to calculate their WCETs. Due to page limitation in this paper, the discussion about partitioned scheduling with cache space isolation is also presented in our technical report [184].

8.9 Conclusions

The broad introduction of multicores brings us many interesting research challenges for embedded systems design. One of these is to predict the timing properties of embedded software on such platforms. One of the main obstacles is the sharing of on-chip caches such as L2. The message of this chapter is that with proper resource isolation, it is possible to perform system-level schedulability analysis for multi-core systems based on task-level timing analysis using existing WCET analysis techniques. Our contributions include two efficient techniques for such analyses in the presence of a shared cache. We may argue that hard real-time applications should be placed in local caches such as L1. An interesting future work is to develop techniques for estimating the cache space requirements of tasks.

However, when there is not enough local cache space, the techniques presented here will be needed. We believe that our analysis techniques are also applicable to handle other types of on-chip resources such as bus bandwidth. We leave this for future work. As future work, we will also study how the allocation of cache space size for individual tasks will influence system-level performance and timing properties.

Appendix: Improving the Interference Computation

The computation of I_k^i, an upper bound of the interference caused by τ_i over J_k, in Eq. (8.3) (Sect. 8.5.1) is grossly over-pessimistic. In the following we will present a more precise computation of I_k^i by carefully identifying the worst-case scenario of τ_i's interference.

Recall that the problem window $[r_k, l_k]$ is a time frame of a given length $(l_k - r_k = S_k)$ for which we want to derive a bound of how much interference a task τ_i (or rather its jobs) can cause to possibly prevent J_k from running. We can compute I_k^i using the following lemma:

Lemma 8.2. *An upper bound of the interference contributed by τ_i in the problem window of length S_k can be computed by*

$$I_k^i = \begin{cases} S_k & i < k \land S_k < C_i \\ \left\lfloor \frac{S_k - C_i}{T_i} \right\rfloor C_i + C_i + \omega & i < k \land S_k \geq C_i \\ 0 & i = k \\ \min(C_i, S_k) & i > k \end{cases} \qquad (8.10)$$

where

$$\omega = \min \left(C_i, \max \left(0, (S_k - C_i) \bmod T_i - (T_i - D_i) \right) \right) \qquad (8.11)$$

Proof. The lemma is proved in the following cases:

1. $i < k$, i.e., τ_i's priority is higher than τ_k's.
   If $S_k < C_i$, i.e., a job of τ_i can execute even longer than J_k's slack, trivially $I_k^i = S_k$ is a safe bound.
   If $S_k \geq C_i$, the worst-case for I_k^i occurs when

   (a) one of τ_i's jobs is released at $l_k - C_k$,
   (b) all jobs are released with period T_i, and
   (c) the carry-in job executes as late as possible.

   See Fig. 8.8. To see that this is indeed the worst-case, we imagine to move the release times of τ_i's jobs leftwards for a distance $\epsilon^l < T_i - C_i$ or rightwards for a distance $\epsilon^r < C_i$, to see if it is possible to increase I_k^i by doing so. (It's easy to see that moving τ_i's jobs' releases more in either direction creates a situation equivalent to one of these two cases. Further, I_k^i cannot be increased if the number of τ_i's jobs in $[r_k, l_k]$ is decreased, which means we only need to consider the scenario that all jobs are released periodically.) If it is moved leftwards by ϵ^l, τ_i's interference cannot increase at neither the left nor the right end of the interval $[r_k, l_k]$, so moving leftwards for a distance $\epsilon^l < T_i - C_i$ will not increase the interference. On the other hand, when moving rightwards by ϵ^r, the interference is increased by no more than ϵ^r at the left end, but decreased by ϵ^r at the right end, so moving rightwards for a distance $\epsilon^r < C_i$ will also not increase the interference. In summary, based on the scenario in Fig. 8.8, I_k^i cannot be increased no matter how we move the release time of τ_i. With this worst-case scenario, we can see that the interference contributed by the carry-out job is C_i, the number of the body jobs is $\lfloor (S_k - C_i)/T_i \rfloor$ (each contributing C_i interference),

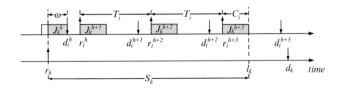

Fig. 8.8 Computation of I_k^i if $i < k$ and $S_k \geq C_i$

and the interference contributed by the carry-in job is bounded by both C_i and the distance between r_k and the carry-in job's deadline. Thus, for each task τ_i with $i < k \wedge S_k \geq C_i$, we can compute I_k^i by

$$I_k^i = \left\lfloor \frac{S_k - C_i}{T_i} \right\rfloor C_i + C_i + \omega \qquad (8.12)$$

where ω is defined as in Eq. (8.11).

2. $i = k$, i.e., τ_i is the analyzed task. Since $D_k \leq T_k$ holds for each task τ_k, the other jobs of τ_k cannot interfere with J_k, so in this case we have

$$I_k^i = 0 \qquad (8.13)$$

3. $i > k$, i.e., τ_i's priority is lower than τ_k.

   In FP_{CA}, a job J_i^h with lower priority than J_k can interfere with J_k only if it is released earlier than r_k. Therefore, τ_i can only cause interference to J_k with at most one job, so its interference is bounded by C_i. The interference is also bounded by the length of the problem window S_k. Thus, for $i > k$, we can compute I_k^i by

$$I_k^i = \min(C_i, S_k) \qquad (8.14)$$

Part III
Real-Time Calculus

Chapter 9
Finitary Real-Time Calculus

Real-time calculus (RTC) is a powerful framework to analyze real-time performance of distributed embedded systems. However, RTC may run into serious analysis efficiency problems when applied to systems of large scale and/or with complex timing parameter characteristics. The main reason is that many RTC operations generate curves with periods equal to the hyper-period of the input curves. Therefore, the analysis in RTC has exponential complexity. In practice the curve periods may explode rapidly when several components are serially connected, which leads to low analysis efficiency.

In this work, we propose *Finitary RTC* to solve the above problem. Finitary RTC only maintains and operates on a limited part of each curve that is relevant to the final analysis results, which results in pseudo-polynomial computational complexity. Experiments show that Finitary RTC can drastically improve the analysis efficiency over the original RTC. The original RTC may take hours or even days to analyze systems with complex timing characteristics, but Finitary RTC typically can complete the analysis in seconds. Even for simple systems, Finitary RTC also typically speeds up the analysis procedure by hundreds of times. While getting better efficiency, Finitary RTC does *not* introduce any extra pessimism, i.e., it yields analysis results as precise as the original RTC.

9.1 Introduction

RTC [186] is a framework for performance analysis of distributed embedded systems rooted in the Network Calculus theory [187]. RTC uses *variability characterization curves* [188] to model workload and resource, and analyzes workload flows through a network of computation and communication resources to derive performance characteristics of the system. RTC has proved to be one of the most

© Springer International Publishing Switzerland 2016
N. Guan, *Techniques for Building Timing-Predictable Embedded Systems*,
DOI 10.1007/978-3-319-27198-9_9

powerful methods for real-time embedded system performance analysis, and has drawn considerable attention from both academia and industry in recent years.

Although one of RTC's advantages is avoiding the state-space explosion problem in state-based verification techniques such as timed automata [189], it may still encounter scalability problems when applied to systems with complex timing properties. The major reason of RTC's efficiency problem is due to the fact that the workload/resource representations and computations in the RTC framework are defined for the infinite range of time intervals. Although in practice it is possible to compactly represent the workload and resource information of infinite time intervals by finite data structures [78], it may still have to maintain a great amount of information in the analysis procedure and take very long time to complete the computations when the timing characteristics of the system are complex. More specifically, many RTC operations generate curves with periods equal to the LCM (least common multiple) of the input curve periods. Therefore, the curve periods may explode rapidly when several components with complex timing characteristics, (e.g., co-prime periods) are connected in a row. We call this phenomenon *period explosion* (Sect. 9.3 discusses the period explosion problem in detail). In general, the analysis in RTC has *exponential* complexity and in practice the efficiency may be extremely low for large systems with complex timing characteristics.

In this chapter, we present *Finitary RTC*, a refinement of the RTC framework, to solve the above problem. Finitary RTC has *pseudo-polynomial* complexity, and in practice can drastically improve the analysis efficiency of complex distributed embedded systems. The key idea is that only the system behavior in time intervals up to a certain length is relevant to the final analysis results, hence we can safely chop off the part of a workload and resource curve beyond that length limit and only work with a (small) piece of the original curve during the analysis procedure.

While getting better efficiency, Finitary RTC does not introduce any extra pessimism, i.e., the analysis results obtained by Finitary RTC are as precise as using the original RTC approach. This is a fundamental difference between Finitary RTC and other methods that approximate the workload and resource information to simplify the problem but lead to pessimistic analysis results (e.g., adjusting task periods for a smaller hyper-period).

We conduct experiments to evaluate the efficiency improvement of Finitary RTC over the original RTC. Experiments show that Finitary RTC can drastically speed up the analysis. For systems with complex timing characteristics, the analysis procedure in the original RTC may take hours or even days, but Finitary RTC typically can complete the analysis in seconds. Even for simple systems, Finitary RTC also typically speeds up the analysis procedure by hundreds of times.

9.1.1 Related Work

RTC is based on Network calculus (NC) [187]. There are several significant differences between RTC and NC. First, RTC can model and compute the remaining service of each component, which is not explicitly considered in NC. Second, while NC mainly uses the upper arrival curves and lower service curves, RTC uses both lower and upper curves for both events and resource, which supports a tighter computation of the output curves. Furthermore, RTC has been extended to model and analyze common problems in the real-time embedded system domain, such as structured event streams [190], workload correlations [191], interface-based design [192], mode switches [193], and cyclic systems [79].

The classical real-time scheduling theory is extended to distributed systems by *holistic analysis* [194]. MAST [195] is a well-known example of this approach. MAST supports offset-based holistic schedulability analysis to guarantee various real-time performance constraints such as local deadlines, global deadlines, and maximal jitters. Different from the holistic analysis approach in MAST, RTC supports a compositional analysis framework, where local analysis is performed for each component to derive the output event/resource models and afterwards the calculated output event models are propagated to the subsequent components. SymTA/S [40] is another well-known compositional real-time performance analysis framework, which uses a similar event and resource model with RTC. The local analysis in SymTA/S is based on the classical busy period technique in real-time scheduling theory [35]. We refer to [196] for comparisons among these frameworks.

State-based verification techniques such as Timed Automata [189] provide extremely powerful expressiveness to model complex real-time systems. However, this approach usually suffers from the state-space explosion problem. The analytical (stateless) method of RTC depends on solutions of closed-form expressions, which in general yields a much better scalability than the state-based approach. Hybrid methods combining RTC and timed automata [197] are used to balance the expressiveness and scalability in the design of complex systems. However, although RTC is significantly more scalable than state-based verification techniques, it still may run into serious efficiency problems due to the *period explosion* phenomenon.

9.2 RTC Basics

9.2.1 Arrival and Service Curves

RTC uses *variability characterization curves* (*curves* for short) to describe timing properties of event streams and available resource:

Definition 9.1 (Arrival Curve). Let $R[s, t)$ denote the total amount of requested capacity to process in time interval $[s, t)$. Then, the corresponding upper and lower arrival curves are denoted as α^u and α^l, respectively, and satisfy:

$$\forall s < t, \ \alpha^l(t - s) \leq R[s, t) \leq \alpha^u(t - s) \tag{9.1}$$

where $\alpha^u(0) = \alpha^l(0) = 0$.

Definition 9.2 (Service Curve). Let $C[s, t)$ denote the number of events that a resource can process in time interval $[s, t)$. Then, the corresponding upper and lower service curves are denoted as β^u and β^l, respectively, and satisfy:

$$\forall s < t, \ \ \beta^l(t - s) \le C[s, t) \le \beta^u(t - s) \tag{9.2}$$

where $\beta^u(0) = \beta^l(0) = 0$.

The arrival and service curves are monotonically non-decreasing. Further, we only consider curves that are piece-wise linear and the length of each linear segment is lower bounded by a constant. Therefore, we exclude curves that are "infinitely complex." The number of linear segments contained by a curve in an interval is polynomial with respect to the interval length. To simplify the presentation, we sometimes use a curve pair α (β) to represent both the upper curve α^u (β^u) and the lower curve α^l (β^l).

The RTC framework intensively uses the min-plus/max-plus convolution and deconvolution operations:

Definition 9.3 (Convolution and Deconvolution). Let f, g be two curves, the min-plus convolution $\otimes$, max-plus convolution $\overline{\otimes}$, min-plus deconvolution $\oslash$, and max-plus deconvolution $\overline{\oslash}$ are defined as

$$(f \otimes g)(\Delta) \triangleq \inf_{0 \le \lambda \le \Delta} \{f(\Delta - \lambda) + g(\lambda)\}$$

$$(f \overline{\otimes} g)(\Delta) \triangleq \sup_{0 \le \lambda \le \Delta} \{f(\Delta - \lambda) + g(\lambda)\}$$

$$(f \oslash g)(\Delta) \triangleq \sup_{\lambda \ge 0} \{f(\Delta + \lambda) - g(\lambda)\}$$

$$(f \overline{\oslash} g)(\Delta) \triangleq \inf_{\lambda \ge 0} \{f(\Delta + \lambda) - g(\lambda)\}$$

We assume an upper (lower) bound curve to be *sub-additive* (*super-additive*) [78]:

Definition 9.4 (Sub-Additivity and Super-Additivity). A curve f is sub-additive iff

$$\forall x, y \ge 0 : f(x) + f(y) \ge f(x + y)$$

A curve f is super-additive iff

$$\forall x, y \ge 0 : f(x) + f(y) \le f(x + y)$$

Moreover, we assume the arrival curve $\alpha = (\alpha^u, \alpha^l)$ and service curves $\beta = (\beta^u, \beta^l)$ to be *causal* [198]:

Definition 9.5 (Causality). Given a sub-additive upper bound curve f^u and a super-additive lower bound curve f^l, curve pair $f = (f^u, f^l)$ is *causal* iff

$$f^u = f^u \overline{\oslash} f^l \quad \text{and} \quad f^l = f^l \oslash f^u$$

The arrival and service curves are *well defined* if they comply with the monotonicity, sub/super-additivity, and causality constraints. Arrival and service curves that are not well-defined contain inconsistent information in modeling realistic system timing behaviors, and can be transferred into their monotonicity, sub/super-additivity and causality closures [78, 198]. The computations in RTC generate well-defined arrival and service curves if the inputs are well defined. We assume that all arrival and service curves are well defined, which is necessary to establish the Finitary RTC. Finally, we define the *long-term slope* of a curve f as

$$s(f) \triangleq \lim_{\Delta \to +\infty} (f(\Delta)/\Delta)$$

9.2.2   Greedy Processing Component

We focus on the most widely used abstract component in RTC called *Greedy processing component* (GPC) [74]. A GPC processes events from the input event stream (described by arrival curve α) in a greedy fashion, as long as it complies with the availability of resources (described by the service curve β). GPC produces an output event stream, described by arrival curve $\alpha' = (\alpha^{u'}, \alpha^{l'})$, and an output remaining service, described by service curve $\beta' = (\beta^{u'}, \beta^{l'})$:

$$\alpha^{u'} \triangleq [(\alpha^u \otimes \beta^u) \oslash \beta^l] \wedge \beta^u \tag{9.3}$$

$$\alpha^{l'} \triangleq [(\alpha^l \oslash \beta^u) \otimes \beta^l] \wedge \beta^l \tag{9.4}$$

$$\beta^{u'} \triangleq (\beta^u - \alpha^l) \overline{\oslash} 0 \tag{9.5}$$

$$\beta^{l'} \triangleq (\beta^l - \alpha^u) \overline{\otimes} 0 \tag{9.6}$$

where $(f \wedge g)(\Delta) = \min(f(\Delta), g(\Delta))$. We use the following form to compactly represent the computation in (9.3) $\sim$ (9.6):

$$(\alpha', \beta') = \mathsf{GPC}(\alpha, \beta)$$

The number of events in the input queue, i.e., the *backlog*, and the *delay* of each event can be bounded by $B(\alpha^u, \beta^l)$ and $D(\alpha^u, \beta^l)$, respectively:

$$B(\alpha^u, \beta^l) \triangleq \sup_{\lambda \geq 0} \{\alpha^u(\lambda) - \beta^l(\lambda)\} \tag{9.7}$$

$$D(\alpha^u, \beta^l) \triangleq \sup_{\lambda \geq 0} \{\inf\{\tau \in [0, \lambda] : \alpha^u(\lambda - \tau) \leq \beta^l(\lambda)\}\} \tag{9.8}$$

Fig. 9.1 Illustration of $B(\alpha^u, \beta^l)$, $D(\alpha^u, \beta^l)$, and $\text{MBS}(\alpha^u, \beta^l)$

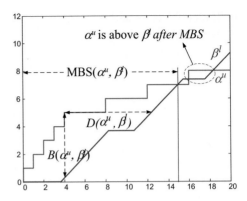

Intuitively, $B(\alpha^u, \beta^l)$ and $D(\alpha^u, \beta^l)$ are the maximal vertical and horizontal distance from α^u to β^l, as illustrated in Fig. 9.1. In this chapter, we intensively use another notation *maximal busy-period size*:

$$\text{MBS}(\alpha^u, \beta^l) = \min\{\lambda > 0 : \alpha^u(\lambda) = \beta^l(\lambda)\}$$

Intuitively, $\text{MBS}(\alpha^u, \beta^l)$ is the maximal length of the time interval in which α^u is above β^l, i.e., the maximal size of the so-called *busy periods* [30] (formally defined in the appendix). Note that in general α^u may go above β^l again after they intersect at $\text{MBS}(\alpha^u, \beta^l)$, as shown in Fig. 9.1. We use MBS as the abbreviation of $\text{MBS}(\alpha^u, \beta^l)$ when α^u and β^l are clear from the context.

We assume for each component $s(\alpha^u)/s(\beta^l)$ is bounded by a constant ε that is strictly smaller than 1, and thus $\text{MBS}(\alpha^u, \beta^l)$ is bounded by a number that is *pseudo-polynomially* large. This is essentially the same as the common constraint in real-time scheduling theory that the system *utilization* is strictly smaller than 1 [199].

9.2.3 Performance Analysis Network

The RTC framework connects multiple components into a network to model systems with resource sharing and networked structures, as illustrated in Fig. 9.2. As in most literature on RTC we assume the performance analysis networks are *acyclic*. Therefore, one can conduct the analysis of the whole network following the event and resource flows: it starts with a number of *initial input curves*, to generate *intermediate curves* step by step, and eventually generate the *final output curves*. The components whose inputs are all initial input curves are called *start components*, and the ones whose outputs are all final output curves are called *end components*.

For example, in Fig. 9.2 the analysis starts with the start component 1, using the initial input curves α_1, β_1 to derive its backlog bound $B(\alpha_1^u, \beta_1^l)$ and delay bound

Fig. 9.2 An example of analysis network in RTC

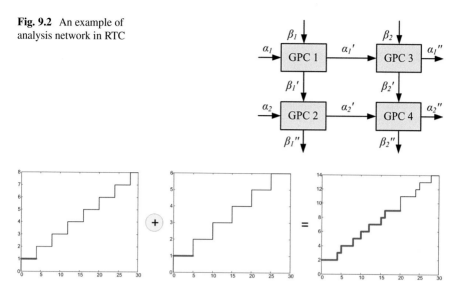

Fig. 9.3 Illustration of the segment number increase caused by a plus operation, where the *red* part of each curve are the segments that need to be stored to represent the whole curve

$D(\alpha_1^u, \beta_1^l)$, and generate the output curves α_1', β_1'. Then α_2, β_1' are used to analyze component 2, and α_2, β_1' are used to analyze component 3, and finally the resulting curves α_2', β_2' from these two components are used to analyze the end component 4.

9.3 Efficiency Bottleneck of RTC

In RTC, the arrival/service curves are defined for the infinite range of positive real numbers $\Delta \in \mathbb{R}_{\geq 0}$. For a practical implementation, we need a finite representation of curves and the curve operations should be completed in a finite time.

To solve this problem, RTC Toolbox [80] restricts to a class of *regular* curves [78], which can be efficiently represented by *finite* data structures but are still expressive enough to model most realistic problems. A regular curve consists of an aperiodic part, followed by a periodic part. Each part is represented by a concatenation of linear *segments*. Generally, the computation time and memory requirement of an operation between two curves is proportional to the number of segments contained by the curves.

Many RTC operations (e.g., plus, minus, and convolution) generate output curves with much longer periods than the periods of the input curves. Typically, the period of the output curve equals to the LCM of the input periods. Figure 9.3 shows an example of the plus operation of two curves. Both input curves are strictly periodic (i.e., the aperiodic parts are empty). The first input curve's period is 4 and the second input curve's period is 5. Each of them only contains one segment. Applying the plus operation to these two curves, the result is a periodic curve with period 20, which is the LCM of 4 and 5, and containing eight segments.

In general, when many components are serially connected in a row, the number of segments contained by the curves increases *exponentially*, and thus the time cost of the analysis increases *exponentially* as it steps forward along the analysis flow. We call this phenomenon *period explosion*, which is the major reason why the analysis of large-scale systems in RTC is problematic.

9.4 Overview of Finitary RTC

We start with a simple example of only one component. As introduced in Sect. 9.2.2, the backlog bound $B(\alpha^u, \beta^l)$ and delay bound $D(\alpha^u, \beta^l)$ of this system are calculated by measuring the maximal horizontal and vertical distance between α^u and β^l (in the part where α^u is above β^l).

Actually, to calculate $B(\alpha^u, \beta^l)$ and $D(\alpha^u, \beta^l)$, one only needs to check both curves up to $\mathsf{MBS}(\alpha^u, \beta^l)$ on the x-axis:

Theorem 9.1. *Given a sub-additive upper arrival curve α^u and a super-additive service curve β^l,*

$$B(\alpha^u, \beta^l) = \sup_{MBS \geq \lambda \geq 0} \{\alpha^u(\lambda) - \beta^l(\lambda)\}$$

$$D(\alpha^u, \beta^l) = \sup_{MBS \geq \lambda \geq 0} \{\inf\{\tau \in [0, \lambda] : \alpha^u(\lambda - \tau) \leq \beta^l(\lambda)\}\}$$

Theorem 9.1 is proved in the appendix.

Note that α^u may be above β^l again after MBS, as shown in Fig. 9.1. Nevertheless, Theorem 9.1 guarantees that the maximal backlog and delay occurs in time intervals of size up to $\mathsf{MBS}(\alpha^u, \beta^l)$. Therefore, the analysis can "chop off" the parts beyond $\mathsf{MBS}(\alpha^u, \beta^l)$ of both curves, and only use the remaining *finitary curves* to obtain exactly the same $B(\alpha^u, \beta^l)$ and $D(\alpha^u, \beta^l)$ results as before. This represents the basic idea of Finitary RTC:

Main Idea: Instead of working with complete curves, we only work with the part of each curve that is relevant to the analysis results.

This is similar to the standard analysis techniques based on *busy periods* in classical real-time scheduling theory. For example, in the schedulability analysis of EDF based on *demand bound functions* [30], one only needs to check what happens in a busy period. The analysis of time intervals beyond the maximal busy period size is irrelevant to the system schedulability and thereby can be ignored.

However, it is difficult to apply this idea to the analysis of networked systems in the RTC framework. For example, suppose we want to analyze the backlog and delay bound of component 4 in Fig. 9.2. By the discussions above, we only need to keep the input curves of component 4 up to $\mathsf{MBS}(\alpha_2^{u'}, \beta_2^{l'})$ on the x-axis to obtain the desired results. However, the value of $\mathsf{MBS}(\alpha_2^{u'}, \beta_2^{l'})$ is not revealed until we have actually obtained $\alpha_2^{u'}$ and $\beta_2^{l'}$. To calculate $\alpha_2^{u'}$ and $\beta_2^{l'}$, we need to first accomplish the analysis of components 2 and 3, and component 1 at the first place. Therefore, we still have to conduct the expensive analysis for the preceding components of component 4 with complete curves, although we only need a small part of the input curves of component 4 to calculate its backlog and delay bound. In this way, the efficiency improvement is trivial since the analysis procedure is as expensive as in the original RTC except for the very last step to analyze component 4.

The target of Finitary RTC is to work with finitary curves (the parts of curves up to certain limits) through the whole analysis network. In other words, we should already use finitary curves at the initial inputs, and generate finitary curves at outputs from the input finitary curves at each component. By this, the overall analysis procedure is significantly more efficient than the original RTC approach. To achieve this, we first need to solve the following problem:

Problem 1: *How to compute the output finitary curves from the input finitary curves for each component?*

Recall that the computation in $(\alpha', \beta') = \mathsf{GPC}(\alpha, \beta)$ uses the min-plus and max-plus deconvolution operations:

$$(f \oslash g)(\Delta) = \sup_{\lambda \geq 0}\{f(\Delta + \lambda) - g(\lambda)\}$$

$$(f \overline{\oslash} g)(\Delta) = \inf_{\lambda \geq 0}\{f(\Delta + \lambda) - g(\lambda)\}$$

In order to calculate $(f \oslash g)(\Delta)$ or $(f \overline{\oslash} g)(\Delta)$ for a particular Δ, it is required to check the value of $f(\Delta + \lambda) - g(\lambda)$ for *all* $\lambda \geq 0$, i.e., slide over curve f from Δ to above and slide over the whole curve g, to get the suprema (infima). Therefore, even if we only want to calculate a small piece of a curve at the output, we still need to know the *complete* input curves defined in the infinite range. Section 9.5 is dedicated to the solution of this problem, where we use a "finitary" version of the deconvolution operations in the computation of the output arrival and service curves. We prove that to calculate the output curve up to interval size x, it is enough to only visit the part of input curves up to interval length $x + \mathsf{MBS}$.

If we somehow know the MBS value for each component, then we can "backtrack" the whole analysis network to decide the size of the input finitary curves for each component, and eventually decide the size of the curves we need to keep at the initial inputs. However, the MBS value for each component is not revealed until its input arrival and service curves are actually known, and we do not want to actually perform the expensive analysis with complete curves to obtain the MBS information. Therefore, we need to solve the following problem:

Problem 2: *How to efficiently estimate the* **MBS** *value of each component in the network?*

We address this problem in Sect. 9.6, using safe approximations of the input curves to quickly "pre-analyze" the whole analysis network and obtain safe estimation of the **MBS** value for each component. The key point is that as long as we always use *over-approximations* (formally defined in Sect. 9.6) of the curves, the obtained **MBS** estimation is guaranteed to be no smaller than its real value. Note that the *over-approximations* of input curves are used merely for the purpose of estimating **MBS**. After we have obtained the **MBS** estimation of each component, the following analysis procedure does not introduce any extra pessimism, and the analysis results we finally obtained are as precise as using the original RTC.

9.5 Analyzing GPC with Finitary Deconvolution

This section addresses the first problem, i.e., how to compute the output finitary curves from the input finitary curves for each component. We first define the finitary version of the min-plus and max-plus deconvolutions operations:

Definition 9.6 (Finitary Deconvolution). The *finitary min-plus deconvolution* and *finitary max-plus deconvolution* regarding a non-negative real number T, denoted by $\oslash_T$ and $\overline{\oslash}_T$ respectively, are defined as

$$(f \oslash_T g)(\Delta) \triangleq \sup_{T \geq \lambda \geq 0} \{f(\Delta + \lambda) - g(\lambda)\}$$

$$(f \overline{\oslash}_T g)(\Delta) \triangleq \inf_{T \geq \lambda \geq 0} \{f(\Delta + \lambda) - g(\lambda)\}$$

The result of $(f \oslash_T g)(\Delta)$ and $(f \overline{\oslash}_T g)(\Delta)$ for a particular Δ only depends on f in $[\Delta, \Delta + T]$ and g in $[0, T]$.

Now we can refine the computation $(\alpha', \beta') = \mathsf{GPC}(\alpha, \beta)$, using $\oslash_T$ and $\overline{\oslash}_T$ to replace $\oslash$ and $\overline{\oslash}$, and still safely bound the output event stream and remaining service:

Theorem 9.2. *Given an event stream described by the arrival curves α^u, α^l and a resource described by the service curves β^u, β^l. If T is a real number with*

$$T \geq \mathsf{MBS}(\alpha^u, \beta^l)$$

then the processed event stream and remaining service are bounded from above and below by $(\alpha', \beta') = \mathsf{GPC}_T(\alpha, \beta)$:

$$\alpha^{u'} \triangleq ((\alpha^u \otimes \beta^u) \oslash_T \beta^l \wedge \beta^u) \tag{9.9}$$

$$\alpha^{l'} \triangleq ((\alpha^l \oslash_T \beta^u) \otimes \beta^l \wedge \beta^l) \tag{9.10}$$

$$\beta^{u'} \triangleq (\beta^u - \alpha^l) \,\overline{\oslash}_T\, 0 \tag{9.11}$$

$$\beta^{l'} \triangleq (\beta^l - \alpha^u) \,\overline{\otimes}\, 0 \tag{9.12}$$

Moreover, GPC_T does *not* sacrifice any analysis precision comparing with GPC that uses the original deconvolution operations:

Theorem 9.3. *The output event arrival curves and remaining service curves obtained by* GPC_T *in Theorem 9.2 is at least as precise as that obtained by* GPC.

The proofs of Theorems 9.2 and 9.3 are presented in the appendix. Note that these proofs are *not* simple reproductions of their counterparts in the original RTC. Particularly, sophisticated techniques are developed to utilize the busy period concept in the construction of the desired bounds in Theorem 9.2, and Theorem 9.3 relies on the sub/super-additivity and causality property of the input arrival and service curves.

The original deconvolution in GPC is replaced by the finitary deconvolution operations in GPC_T, so the computation of a particular point on the output curve only requires to slide over a limited range of the input curves. Therefore, we can establish the information dependency between the input curves and output curves in GPC_T:

Theorem 9.4. *Let* $(\alpha', \beta') = \mathsf{GPC}_T(\alpha, \beta)$. *The computation of* α' *or* β' *in the range of* $[0, x]$ *on the x-axis only depends on the input curves in the range of* $[0, x+T]$ *on the x-axis.*

Theorem 9.4 can be easily proved by rewriting (9.9) $\sim$ (9.12) with the definition of the convolution and finitary deconvolution operations. Proof details are omitted due to space limit.

We use $|f|$ to denote the *upper limit* on the x-axis to which we want to keep the curve (pair) f. Then according to Theorem 9.4, we know the following constraint between the upper limits of the input and output curves in GPC_T:

$$|\alpha| = |\beta| \geq T + \max(|\alpha'|, |\beta'|) \tag{9.13}$$

For example, if we want to use GPC_6 to generate output curves with upper limit of 5, then the upper limit of the input curves should be at least $5 + 6 = 11$, i.e., the input curves should be defined at least in the range $[0, 11]$.

9.6 Analysis Network in Finitary RTC

The GPC networks considered in this work are *acyclic*. Therefore, as soon as we have chosen the T value in GPC_T for each component, we can traverse the network backwards and use the relation in (9.13) to iteratively decide the upper limits of the input finitary curves of each component. The choice of T for each component is subject to the constraint $T \geq$ MBS. Therefore, it is sufficient to derive an upper bound of MBS, and use this upper bound as the value of T in GPC_T at each component. In the following, we first introduce how to efficiently compute a safe upper bound of MBS_i for each component i, then introduce how to decide the upper limits of input finitary curves of each component.

9.6.1   Bounding Individual MBS

We first define the over-approximation of curves:

Definition 9.7 (Over-Approximation). For two upper bound curves c^{u*} and c^u, if $\forall \Delta \geq 0 : c^{u*}(\Delta) \geq c^u(\Delta)$, then c^{u*} is an over-approximation of c^u. For two lower bound curves c^{l*} and c^l, if $\forall \Delta \geq 0 : c^{l*}(\Delta) \leq c^l(\Delta)$, then c^{l*} is an over-approximation of c^l. Curve pair $c^* = (c^{u*}, c^{l*})$ is an over-approximation of curve pair $c = (c^u, c^l)$ if c^{u*} and c^{l*} are over-approximations of c^u and c^l, respectively. We use $a \succeq b$ to denote that a is an over-approximation of b.

To compute a safe upper bound of MBS_i for each component, we use over-approximations of the initial input curves to "pre-analyze" the whole network. First we have the following property by examining the computation rules of GPC:

Property 9.1. If $\alpha_1 \succeq \alpha_2$ and $\beta_1 \succeq \beta_2$, then $\alpha'_1 \succeq \alpha'_2$ and $\beta'_1 \succeq \beta'_2$, where $(\alpha'_1, \beta'_1) = \mathsf{GPC}(\alpha_1, \beta_1)$ and $(\alpha'_2, \beta'_2) = \mathsf{GPC}(\alpha_2, \beta_2)$.

So we know that if we start with over-approximations of the initial input curves, then during the whole "pre-analysis" procedure all the resulting curves are over-approximations of their correspondences. Further, we know

Property 9.2. If $\alpha^u_1 \succeq \alpha^u_2$ and $\beta^l_1 \succeq \beta^l_2$, then $\mathsf{MBS}(\alpha^u_1, \beta^l_1) \geq \mathsf{MBS}(\alpha^u_2, \beta^l_2)$.

Therefore, if we use over-approximations of the initial input curves to conduct the analysis of the system, then the resulted maximal busy-period size of each component i, denoted by MBS^*_i, is a safe upper bound of the real MBS_i of that component with the original curves.

There are infinitely many possibilities to over-approximate the initial input curves to conduct the pre-analysis procedure. We choose to use the "tightest" linear functions as their over-approximations: each initial input curve pair $f = (f^u, f^l)$ is over-approximated by $\overline{f} = (\overline{f^u}, \overline{f^l})$ where $\overline{f^u}(\Delta) = a^u \times \Delta + b^u$ and $\overline{f^l}(\Delta) = a^l \times \Delta + b^l$ with

$$a^u = s(f^u) \quad b^u = \inf\{\, b \mid \forall \Delta : a^u \times \Delta + b \geq f^u(\Delta)\}$$
$$a^l = s(f^l) \quad b^l = \sup\{\, b \mid \forall \Delta : a^l \times \Delta + b \leq f^l(\Delta)\}$$

Using these linear functions as the initial inputs, the generated curves in the pre-analysis procedure only contain polynomially many segments and the computations are very efficient.

9.6.2 Decision of Curve Upper Limits

The decision of the upper limit of each finitary curve starts with the end components. Since the output curves of an end component are not used by other components, it is enough to set the upper limit of the input finitary curves of each end component i to be MBS_i^*. Then we can traverse the whole analysis network backwards, using (9.13) to iteratively compute the upper limits of the finitary input curves of each component. The pseudo-code of this procedure is shown in Algorithm 5.

Example 9.1. We use linear functions to over-approximations the initial inputs $\alpha_1, \alpha_2, \beta_1, \beta_2$ to pre-analyze the system in Fig. 9.2. Suppose the resulting estimated maximal busy-period sizes are $\mathsf{MBS}_1^* = 10$, $\mathsf{MBS}_2^* = 12$, $\mathsf{MBS}_3^* = 14$, and $\mathsf{MBS}_4^* = 20$. Then we iteratively compute the size of the finitary curves:

$$|\alpha_2'| = |\beta_2'| = \mathsf{MBS}_4^* = 20$$

$$|\alpha_1'| = |\beta_2| = \mathsf{MBS}_3^* + \max(|\beta_2'|, 0) = 14 + 20 = 34$$

$$|\alpha_2| = |\beta_1'| = \mathsf{MBS}_2^* + \max(|\alpha_2'|, 0) = 12 + 20 = 32$$

$$|\alpha_1| = |\beta_1| = \mathsf{MBS}_1^* + \max(|\alpha_1'|, |\beta_1'|) = 10 + 34 = 44$$

1: Mark all component as *unfinished*.
2: **for** each end component i **do**
3:      $|\alpha_i^{in}| = |\beta_i^{in}| \leftarrow \mathsf{MBS}_i^*$
4:      Mark component i as *finished*
5: **end for**
6: **while** exist *unfinished* components **do**
7:      Select an *unfinished* component j whose successors are both *finished* (i.e., $|\alpha_j^{out}|$ and $|\beta_j^{out}|$ are known).
8:      $|\alpha_j^{in}| = |\beta_j^{in}| \leftarrow \mathsf{MBS}_j^* + \max(|\alpha_j^{out}|, |\beta_j^{out}|)$
9:      Mark component j as *finished*
10: **end while**

Algorithm 5: Pseudo-code of algorithm computing the upper limit of each finitary curve.

9.6.3   Complexity

In summary, the analysis of a network by Finitary RTC consists of three steps: (1) Using linear over-approximations of initial input curves to analyze the network "forwards" and get an estimated MBS^* for each component. (2) Using the estimated MBS^* values to traverse the analysis network "backwards" to decide the upper limits of the input finitary curves of each component. (3) Conduct the analysis of the network "forwards" with the finitary curves.

Finding the "tightest" linear over-approximation of a curve is of pseudo-polynomial complexity (polynomial with respect to the number of segments contained by the curve). With the linear over-approximations of the initial input curves, the maximal number of segments of each curve generated in the pre-analysis is polynomial with respect to network size. So the overall complexity of step (1) is pseudo-polynomial. It is easy to see that the complexity of step (2) is polynomial. Since for each component $s(\alpha^u)/s(\beta^l)$ is bounded by a constant ε strictly smaller than 1, the estimated MBS^* with the linear curve over-approximations is bounded by a number that is pseudo-polynomially large. Therefore, each finitary curve obtained by Algorithm 5 has a pseudo-polynomial upper limit. Since the complexity of GPC_T is polynomial with respect to the curve upper limits and the value of T (i.e., the estimated MBS^*), the complexity of the analysis for each component is pseudo-polynomial, and the overall complexity of step (3) is also pseudo-polynomial. In summary, the overall complexity of the whole analysis procedure is pseudo-polynomial.

9.7   Evaluation

In this section we use synthetic systems to evaluate the efficiency improvement of the Finitary RTC approach. We use the metric *speedup ratio* to represent the efficiency improvement of Finitary RTC over the original RTC:

$$\text{speedup ratio} = \frac{\text{time cost of analysis by original RTC}}{\text{time cost of analysis by Finitary RTC}}$$

In Sect. 9.7.1 we present a case study to give a general picture of the efficiency improvement by Finitary RTC. Then in Sect. 9.7.2 we adjust the parameters of this system to discuss different factors that affect the speedup ratio.

We implement Finitary RTC in RTC Toolbox [80]. RTC Toolbox consists of two major software components: a Java-implemented kernel of basic RTC operations and a set of Matlab libraries to provide high-level modeling capability. Since the Java kernel is not open-source, we implement Finitary RTC only using the open-source Matlab libraries: For each finitary curve f, we use a linear curve to replace the part beyond its upper limit $|f|$ instead of actually removing this part. Therefore,

Fig. 9.4 Analysis network of
the case study

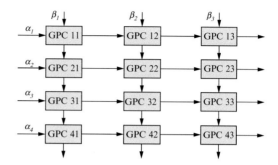

we can reuse the Java kernel of RTC Toolbox to implement the idea of Finitary RTC.
All experiments are conducted on a desktop computer with a 3.4GHZ Intel Core i7
processor.

9.7.1 Case Study

We consider a fairly small and simple system of a 4×3 2D-mesh analysis network, as
shown in Fig. 9.4. All initial input event curves are specified by the parameter triple
(p, j, d), where p denotes the period, j the jitter, and d the minimum inter-arrival
distance of events [200]. The arrival curves of a (p, j, d)-specified stream are

$$\alpha^u(\Delta) = \min \left(\left\lceil \frac{\Delta + j}{p} \right\rceil, \left\lceil \frac{\Delta}{d} \right\rceil \right), \quad \alpha^l(\Delta) = \left\lfloor \frac{\Delta - j}{p} \right\rfloor$$

All initial input service curves correspond to TDMA resource. Each resource is
specified by a triple (s, c, b), where a slot of length s is allocated with very TDMA
cycle c, on a resource with total bandwidth b [80]. The service curves of a (s, c, b)-
specified resource are

$$\beta^u(\Delta) = \left(\left\lfloor \frac{\Delta}{c} \right\rfloor \cdot s + \min(\Delta \bmod c, s) \right) \cdot b$$

$$\beta^l(\Delta) = \left(\left\lfloor \frac{\Delta'}{c} \right\rfloor \cdot s + \min(\Delta' \bmod c, s) \right) \cdot b$$

where $\Delta' = \max(\Delta - c + s, 0)$. The parameters of the input arrival and service
curves are shown in Table 9.1.

Using the original RTC approach, the analysis of the whole network completes
in 415 seconds, while using the Finitary RTC approach, the analysis only takes 0.14
seconds. The Finitary RTC approach leads to a speedup ratio of about 3000 in this
case study.

Table 9.1 Parameters of the initial input arrival and service curves

	α_1	α_2	α_3	α_4
p	10	14	18	22
j	2	3	5	6
d	4	6	8	4

	β_1	β_2	β_3
s	4	6	8
c	6	8	10
b	1	1	1

If we modify the system parameters by changing the period p of the third and fourth event stream to 19 and 23, respectively, which leads to a larger hyper-period of the four event streams, then the original RTC approach cannot accomplish the analysis since an overflow error occurs in the Java kernel of the RTC-Toolbox, while the analysis by Finitary RTC still completes within 0.14 s.

9.7.2 Factors That Affect the Speedup Ratio

The speedup ratio between the Finitary RTC and original RTC heavily depends on the system parameters. In this section we adjust the parameters of the example system in the above case study to discuss factors that affect the speedup ratio.

As introduced in Sect. 9.3, in the original RTC the period of the output curves typically equals the LCM of the periods of the input curves. For example, a system with initial input curves having co-prime periods may lead to very large curve periods. To evaluate the analysis efficiency of the original RTC with different parameter complexity degree, we adjust the period parameter p of each initial input arrival curve to get systems with different hyper-periods. We randomly generate period p of each curve (within a certain scope), to construct systems with different hyper-periods of input event streams, then calculate the average analysis time cost in the original RTC and Finitary RTC of systems with hyper-periods in a certain scope, as shown in Table 9.2. For example, the column starting with "1 ~ 2" in Table 9.2 reports the average analysis time cost in the original RTC and Finitary RTC of systems with hyper-period between 1000 and 2000. From Table 9.2 we can see that the analysis time cost in the original RTC grows rapidly as the hyper-period increases, while Finitary RTC is not sensitive to the hyper-period changes and keeps very low time cost all the time. The speedup ratio is more significant for systems with more complex timing characteristics.

The analysis efficiency of Finitary RTC depends on the curve upper limits. In general, the closer is the long-term ratio of α^u to β^l, the bigger is $\mathsf{MBS}(\alpha^u, \beta^l)$ and thus the bigger is the upper limits of the finitary curves. We define the maximal *utilization* among all the components in the analysis network:

$$U_{max} = \max_{\text{each component } i} \left\{ \frac{s(\alpha_i^u)}{s(\beta_i^l)} \right\}$$

Table 9.2 Experiment results of systems with different hyper-periods

Hyper-period ($\times 10^3$)	1~2	2~3	3~4	4~5	5~6	6~7	7~8	8~9	9~10	10~11	11~12	12~13
Original RTC (second)	2.15	14.51	108.94	195.77	362.95	487.65	643.097	797.32	841.75	1043.68	1342.16	1696.53
Finitary RTC (second)	0.14	0.14	0.14	0.14	0.14	0.14	0.14	0.14	0.14	0.14	0.14	0.14
Speedup ratio	15	112	778	1398	2592	3483	4592	5695	6012	7454	9586	12,118

Table 9.3 Experiment results of systems with different maximal utilizations

b (bandwidth)	1	0.9	0.8	0.7	0.6	0.5	0.45	0.425
U_{max}	0.41	0.45	0.51	0.58	0.68	0.82	0.91	0.96
Original RTC (second)	415.58	487.24	522.38	419.58	388.64	271.01	322.17	305.77
Finitary RTC (second)	0.14	0.14	0.15	0.19	0.20	0.40	14.25	12.74
Speedup ratio	2968	3478	3480	2747	2092	970	19	24

where α_i^u and β_i^l are the input upper arrival and lower service curve of component i. We adjust the bandwidth b for all of the three (s, c, b)-specified TDMA resource in Table 9.1 to construct systems with different U_{max}. We are only interested in systems with $U_{max} \leq 1$, since otherwise some components in the system deem to have unbounded backlog and delay, and the system is considered to be a failure immediately. From Table 9.3 we can see that the time cost of the Finitary RTC approach increases as we increase U_{max}. Nevertheless, Finitary RTC can speed up the analysis by 1000 times with systems of relatively high U_{max} (up to 0.8). Even for systems with U_{max} that are pretty close to 1 (e.g., $U_{max} = 0.96$), Finitary RTC still can significantly improve the analysis efficiency (speed up the analysis by 20 times).

The above experiments show that Finitary RTC drastically speeds up the analysis under different parameter configurations. The speedup is more significantly for systems with more complex timing characteristics and lighter workloads.

9.8 Conclusions and Future Work

In this chapter we present Finitary RTC to drastically improve the analysis efficiency of the RTC framework. The central idea is to use a (small) piece of each infinite curve in the analysis procedure, which avoids the "period explosion" problem in the original RTC and results in a pseudo-polynomial complexity (the original RTC has exponential complexity). In practice, Finitary RTC can drastically improve the analysis efficiency, especially for systems with complex timing characteristics. Finitary RTC does not introduce any extra pessimism, i.e., its analysis results are as precise as using the original RTC.

Although we present Finitary RTC in the context of a particular type of component GPC, the same idea can also be extended to components modeling different scheduling policies like FIFO and EDF. In this work we assume the analysis networks to be *acyclic*. It is not clear yet how to generalize Finitary RTC to the analysis of systems with cyclic event and resource dependencies. A straightforward extension of Finitary RTC to cyclic systems is to bound the number of iterations after which the analysis is guaranteed to converge, and extend the curve upper limits following the analysis iterations, which may lead to very large curve upper limits and less significant efficiency improvement. In the future we will study efficient application of Finitary RTC to cyclic systems.

Appendix 1: Proof of Theorem 9.1

We first prove that in order to compute $B(\alpha^u, \beta^l)$ it is sufficient to only check α^u and β^l up to $\mathsf{MBS}(\alpha^u, \beta^l)$. More precisely, we prove

$$\sup_{\lambda \geq 0} \{\alpha^u(\lambda) - \beta^l(\lambda)\} = \sup_{\mathsf{MBS} \geq \lambda \geq 0} \{\alpha^u(\lambda) - \beta^l(\lambda)\} \qquad (9.14)$$

Let $m = \mathsf{MBS}(\alpha^u, \beta^l)$, so $\alpha^u(m) = \beta^l(m)$. For any non-negative λ, let $q = \lfloor \frac{\lambda}{m} \rfloor$ and $r = \lambda - q \cdot m$, and by the sub-additivity of α^u and super-additivity of β^l we have

$$\alpha^u(\lambda) = \alpha^u(q \cdot m + r) \leq \alpha^u(q \cdot m) + \alpha^u(r) \leq q \cdot \alpha^u(m) + \alpha^u(r)$$
$$\beta^l(\lambda) = \beta^l(q \cdot m + r) \geq \beta^l(q \cdot m) + \beta^l(r) \geq q \cdot \beta^l(m) + \beta^l(r)$$

by which we have

$$\alpha^u(\lambda) - \beta^l(\lambda) \leq q \cdot (\alpha^u(m) - \beta^l(m)) + \alpha^u(r) - \beta^l(r)$$
$$\Leftrightarrow \quad \alpha^u(\lambda) - \beta^l(\lambda) \leq \alpha^u(r) - \beta^l(r) \quad (\because \alpha^u(m) = \beta^l(m))$$

In other words, for any $\lambda \geq 0$, we can always find a corresponding r in the range of $[0, m]$ ($m = \mathsf{MBS}(\alpha^u, \beta^l)$) such that $\alpha^u(\lambda) - \beta^l(\lambda) \leq \alpha^u(r) - \beta^l(r)$, which proves (9.14).

To prove the theorem for $D(\alpha^u, \beta^l)$, we can use the same reasoning as above to the "inverse functions" of $\alpha^u(\lambda)$ and $\beta^l(\lambda)$: $D(\alpha^u, \beta^l)$ is the horizontal distance between the "inverse functions" of $\alpha^u(\lambda)$ and $\beta^l(\lambda)$, and the "inverse function" of $\alpha^u(\lambda)$ is super-additive and the "inverse function" of $\beta^l(\lambda)$ is sub-additive. We omit the proof details for $D(\alpha^u, \beta^l)$ due to space limit.

Appendix 2: Proof of Theorem 9.2

We first introduce some useful concepts and notations [78]: $R[s, t)$ denotes the number of events arrived in time interval $[s, t)$, and $R'[s, t)$ denotes the number of *processed* events in $[s, t)$. $C[s, t)$ denotes the amount of available resource in $[s, t)$, and $C'[s, t)$ denotes the amount of remaining resource in $[s, t)$. Further, the following relation is known [78]:

$$R'[s, t) = C[s, t) - C'[s, t) \qquad (9.15)$$

$B(t)$ denotes the backlog at time t. Moreover, we use p to denote an arbitrarily early time point with $B(p) = 0$. The RTC framework assumes there always exists such a time point p.

Definition 9.8 (Busy Period). A time interval (s, t) is a *busy period* iff both of the following conditions hold: (i) $\forall x \in (s, t) : B(x) \neq 0$ and (ii) $B(s) = B(t) = 0$. Moreover, we call s the *start point* of the busy period (s, t), and t its *end point*.

Lemma 9.1. *Given an event stream restricted by upper arrival curve α^u and resource restricted by lower service curves β^l, the size of any busy period is bounded by $\mathsf{MBS}(\alpha^u, \beta^l)$.*

Proof. Follows the definitions of MBS and busy period.

Then we introduce another important auxiliary lemma:

Lemma 9.2. *Let x_2 be an arbitrary time point with $B(x_2) = 0$. Then $\forall x_1 : p \leq x_1 \leq x_2$:*

$$\sup_{x_1 \leq x \leq x_2} \{C[p, x] - R[p, x]\} = C'[p, x_2] \tag{9.16}$$

Proof. Since $B(x_2)$, all the events arrived before x_2 have been processed before x_2, so we have

$$C'[p, x_2] = C[p, x_2] - R[p, x_2] - B(p)$$
$$= C[p, x_2] - R[p, x_2] \quad (\because B(p) = 0)$$

So in the following we only need to prove

$$C[p, x_2] - R[p, x_2] = \sup_{x_1 \leq x \leq x_2} \{C[p, x] - R[p, x]\} \tag{9.17}$$

It is easy to see that the LHS of (9.17) is no larger than its RHS. In the following we only need to prove that it holds LHS $\geq$ RHS as well. We do this by contradiction, assuming

$$C[p, x_2] - R[p, x_2] < \sup_{x_1 \leq x \leq x_2} \{C[p, x] - R[p, x]\}$$

So there must exist a time point $x' \in [x_1, x_2)$ such that

$$C[p, x_2] - R[p, x_2] < C[p, x'] - R[p, x'] \tag{9.18}$$
$$\Leftrightarrow C[x', x_2] < R[x', x_2] \tag{9.19}$$

i.e., the total service provided in $[x', x_2)$ is strictly smaller than the events arrived in the same time interval, which leads to a contradiction with $B(x_2) = 0$.

In the following we prove that the upper/lower bounds in Theorem 9.2 are safe. More specifically, for any two time points s, t with $t - s = \Delta \geq 0$, we prove the following inequalities:

$$\alpha^{u'} : \ R'[s,t] \leq \min(\sup_{0\leq\lambda\leq T}\{\inf_{0\leq\mu\leq\lambda+\Delta}\{\alpha^{u}(\mu)$$

$$+ \beta^{u}(\lambda+\Delta-\mu)\} - \beta^{l}(\lambda)\}, \beta^{u}(\Delta))$$

$$\alpha^{l'} : \ R'[s,t] \geq \min(\inf_{0\leq\mu\leq\Delta}\{\sup_{0\leq\lambda\leq T}\{\alpha^{l}(\mu)$$

$$- \beta^{u}(\lambda+\Delta-\mu)\} + \beta^{l}(\lambda)\}, \beta^{l}(\Delta))$$

$$\beta^{u'} : \ C'[s,t] \leq \inf_{\Delta\leq\lambda\leq T}\{\beta^{u}(\lambda)-\alpha^{l}(\lambda)\}^{+}$$

$$\beta^{l'} : \ C'[s,t] \geq \sup_{0\leq\lambda\leq\Delta}\{\beta^{l}(\lambda)-\alpha^{u}(\lambda)\}$$

Proof of $\alpha^{u'}$

First of all, by $R'[s,t] = R'[p,t] - R'[p,s]$ and (9.15) we have

$$R'[s,t] = C[p,t] - C'[p,t] - C[p,s] + C'[p,s] \tag{9.20}$$

We define a time point s' as follows:

- If $B(s) = 0$, then let $s' = s$.
- If $B(s) \neq 0$, then let s' be the start point of the busy period containing s. By $T \geq$ MBS and Lemma 9.1 we know $s - T \leq s'$.

Then we can apply Lemma 9.2 ($x_1 = s - T$ and $x_2 = s'$) to get

$$C'[p,s'] = \sup_{s-T\leq b\leq s'}\{C[p,b] - R[p,b]\}$$

Therefore, we can rewrite (9.20) as

$$R'[s,t] = \sup_{s-T\leq b\leq s'}\{C[s,t] - C'[p,t] + C[p,b] - R[p,b]\} \tag{9.21}$$

Now we focus on the expression inside the sup-operation in the above equation, with an arbitrary b satisfying $s - T \leq b \leq s'$.

By the same way as defining s', we define t' with respect to t, and apply Lemma 9.2 ($x_1 = b$ and $x_2 = t'$) to get

$$C'[p,t'] = \sup_{b\leq a\leq t'}\{C[p,a] - R[p,a]\} \tag{9.22}$$

We discuss in two cases

- If $B(t) = 0$, i.e., $t = t'$, we can rewrite (9.22) as

$$C'[p, t'] = \sup_{b \le a \le t} \{C[p, a] - R[p, a]\}$$

- If $B(t) \ne 0$, i.e., t' is the start point of the busy period containing t. Therefore, for any time point $c \in (t', t]$, the available resource in time interval $[t', c)$ is strictly smaller than the request in that interval (otherwise the busy period terminates at c), i.e.,

$$\forall c \in (t', t] : C[t', c) - R[t', c) < 0$$

$$\Leftrightarrow \forall c \in (t', t] : C[p, c) - R[p, c) < C[p, t') - R[p, t')$$

$$\Rightarrow \sup_{b \le a \le t} \{C[p, a) - R[p, a)\} = \sup_{b \le a \le t'} \{C[p, a) - R[p, a)\}$$

Combining this with (9.22) we get

$$C'[p, t'] = \sup_{b \le a \le t} \{C[p, a) - R[p, a)\}$$

In summary, no matter whether $B(t)$ equals 0 or not, we have

$$C'[p, t'] = \sup_{b \le a \le t} \{C[p, a) - R[p, a)\}$$

And by the definition of t' we know $C'[t', t) = 0$, we have

$$C'[p, t) = C'[p, t'] = \sup_{b \le a \le t} \{C[p, a) - R[p, a)\}$$

So we can rewrite (9.21) as

$$R'[s, t] = \sup_{s-T \le b \le s'} \{C[s, t) - \sup_{b \le a \le t} \{C[p, a) - R[p, a)\}$$

$$+ C[p, b) - R[p, b)\}$$

$$= \sup_{s-T \le b \le s'} \{ \inf_{b \le a \le t} \{C[s, t) + C[a, b) - R[a, b)\}\}$$

$$\le \sup_{s-T \le b \le s} \{ \inf_{b \le a \le t} \{C[s, t) + C[a, b) - R[a, b)\}\}$$

We define $\lambda = s - b$ and $\mu = a + \lambda - s$. Since $a \ge b$, we also know $\mu \ge 0$. Further, $\Delta = t - s$. Applying these substitutions to above we have

$$R'[s, t] \le \sup_{0 \le \lambda \le T} \{ \inf_{0 \le \mu \le \lambda + \Delta} \{R[s - \lambda, \mu - \lambda + s)$$

$$+ C[\mu - \lambda + s, t) - C[s - \lambda, s)\}$$

Use the upper and lower curves to substitute R and C in above:

$$R'[s,t] \leq \sup_{0 \leq \lambda \leq T} \{ \inf_{0 \leq \mu \leq \lambda + \Delta} \{ \alpha^u(\mu) + \beta^u(\lambda + \Delta - \mu) \} - \beta^l(\lambda) \}$$

Further, it is obvious that $R'[s,t]$ is also bounded by $\beta^u(\Delta)$, so finally we have

$$R'[s,t] \leq \min(\sup_{0 \leq \lambda \leq T} \{ \inf_{0 \leq \mu \leq \lambda + \Delta} \{ \alpha^u(\mu) + \beta^u(\lambda + \Delta - \mu) \}$$

$$- \beta^l(\lambda) \}, \beta^u(\Delta))$$

Proof of $\alpha^{l'}$

By the computation of $\alpha^{l'}$ in the original **GPC** we have

$$R'[s,t] \geq \min(\inf_{0 \leq \mu \leq t-s} \{ \sup_{0 \leq \lambda} \{ \alpha^l(\mu) - \beta^u(\lambda + t - s - \mu) \}$$

$$+ \beta^l(\lambda) \}, \beta^l(t-s))$$

$$\Rightarrow R'[s,t] \geq \min(\inf_{0 \leq \mu \leq t-s} \{ \sup_{0 \leq \lambda \leq T} \{ \alpha^l(\mu) - \beta^u(\lambda + t - s - \mu) \}$$

$$+ \beta^l(\lambda) \}, \beta^l(t-s))$$

Proof of $\beta^{u'}$

By the computation of $\beta^{u'}$ in the original **GPC** we have

$$C'[s,t] \leq \inf_{t-s \leq \lambda} \{ \beta^u(\lambda) - \alpha^l(\lambda) \}^+$$

$$\Rightarrow C'[s,t] \leq \inf_{t-s \leq \lambda \leq T} \{ \beta^u(\lambda) - \alpha^l(\lambda) \}^+$$

Proof of $\beta^{l'}$

The proof is trivial since the computation of $\beta^{l'}$ in $\mathbf{GPC}_T$ is exactly the same as in the original **GPC**.

Appendix 3: Proof of Theorem 9.3

We first introduce an important lemma:

Lemma 9.3. *Given well-defined arrival and service curves α and β, and two arbitrary non-negative real numbers x and y,*

$$\alpha^l(x) \geq \alpha^l(x+y) - \alpha^u(y) \tag{9.23}$$

$$\beta^u(x) \leq \beta^u(x+y) - \beta^l(y) \tag{9.24}$$

Proof. Since $\alpha = (\alpha^u, \alpha^l)$ is a well-defined (thus *causal*) arrival curve pair, we know $\alpha^l = \alpha^l \oslash \alpha^u$ (see Definition 9.5), i.e.,

$$\alpha^l(x) = \sup_{\lambda \geq 0}\{\alpha^l(x+\lambda) - \alpha^u(\lambda)\}$$

$$\Rightarrow \quad \alpha^l(x) \geq \alpha^l(x+y) - \alpha^u(y)$$

Since $\beta = (\beta^u, \beta^l)$ is also a causal service curve pair, we know $\beta^u = \beta^u \overline{\oslash} \beta^l$, i.e.,

$$\beta^u(x) = \inf_{\lambda \geq 0}\{\beta^u(x+\lambda) - \beta^l(\lambda)\}$$

$$\Rightarrow \quad \beta^u(x) \leq \beta^u(x+y) - \beta^l(y)$$

Now we start to prove Theorem 9.3. More specifically, we shall prove that the output arrival/service upper curves obtained by GPC_T are no larger than its counterpart obtained by GPC, and the output arrival/service lower curves are no smaller than that obtained by GPC.

Proof of $\alpha^{u'}$

Let $F(\Delta, \lambda) = \inf_{0 \leq \mu \leq \lambda + \Delta}\{\alpha^u(\mu) + \beta^u(\lambda + \Delta - \mu)\} - \beta^l(\lambda)$. To prove that the $\alpha^{u'}$ obtained by GPC_T is no larger than that obtained by GPC, we need to show that $\forall \Delta \geq 0$:

$$\min\{\sup_{0 \leq \lambda \leq T}\{F(\Delta, \lambda)\}, \beta^l(\Delta)\} \leq \min\{\sup_{0 \leq \lambda}\{F(\Delta, \lambda)\}, \beta^l(\Delta)\}$$

which is obviously true.

Proof of $\alpha^{l'}$

The main task of this proof is to show

$$\sup_{0 \le \lambda \le T} \{\alpha^l(\mu + \lambda) - \beta^u(\lambda)\} = \sup_{0 \le \lambda} \{\alpha^l(\mu + \lambda) - \beta^u(\lambda)\} \tag{9.25}$$

As soon as (9.25) is proved, it is easy to see that the $\alpha^{l'}$ obtained by GPC_T is no smaller than its counterpart obtained in GPC. In the following we prove (9.25).

For simplicity of presentation, let $m = \mathsf{MBS}(\alpha^u, \beta^l)$, so $\alpha^u(m) = \beta^l(m)$. For an arbitrary λ with $\lambda \ge 0$, let $q = \lfloor \frac{\lambda}{m} \rfloor$ and $r = \lambda - q \cdot m$. Then by Lemma 9.3 and the sub-additivity of α^u we know

$$\alpha^l(\mu + \lambda) - \alpha^l(\mu + r) \le \alpha^u(q \cdot m) \le q \cdot \alpha^u(m) \tag{9.26}$$

On the other hand, by Lemma 9.3 and the super-additivity of β^l we also know

$$\beta^u(\mu + \lambda) - \beta^u(\mu + r) \ge \beta^l(q \cdot m) \ge q \cdot \beta^l(m) \tag{9.27}$$

By (9.26), (9.27), and $\alpha^u(m) = \beta^l(m)$ we can get

$$\alpha^l(\mu + \lambda) - \beta^u(\mu + \lambda) \le \alpha^l(\mu + r) - \beta^u(\mu + r)$$

In other words, for any $\lambda \ge 0$, we can always find a corresponding r in the range of $[0, m]$ such that $\alpha^l(\mu + \lambda) - \beta^u(\mu + \lambda) \le \alpha^l(\mu + r) - \beta^u(\mu + r)$, and by $m \le T$ we finally get (9.25).

Proof of $\beta^{u'}$

We want to prove

$$\inf_{0 \le \lambda \le T} \{\beta^u(\lambda) - \alpha^l(\lambda)\}^+ = \inf_{0 \le \lambda} \{\beta^u(\lambda) - \alpha^l(\lambda)\}^+ \tag{9.28}$$

Similar with the above proof, let $m = \mathsf{MBS}(\alpha^u, \beta^l)$, so $\alpha^u(m) = \beta^l(m)$. Let $q = \lfloor \frac{\lambda}{m} \rfloor$ and $r = \lambda - q \cdot m$. Then by Lemma 9.3 and the sub-additivity of α^u we know

$$\alpha^l(\lambda) - \alpha^l(r) \le \alpha^u(q \cdot m) \le q \cdot \alpha^u(m) \tag{9.29}$$

On the other hand, by Lemma 9.3 and the super-additivity of β^l we can get

$$\beta^u(\lambda) - \beta^u(r) \ge \beta^l(q \cdot m) \ge q \cdot \beta^l(m) \tag{9.30}$$

By (9.29), (9.30) and $\alpha^u(m) = \beta^l(m)$ we have

$$\beta^u(\lambda) - \alpha^l(\lambda) \geq \beta^u(r) - \alpha^l(r)$$

In other words, for any $\lambda \geq 0$, we can always find a corresponding r in the range of $[0, m]$ such that $\beta^u(\lambda) - \alpha^l(\lambda) \geq \beta^u(r) - \alpha^l(r)$, and since $m \leq T$, finally we get (9.28).

Proof of $\beta^{l'}$

The proof is trivial since the computation of $\beta^{l'}$ in GPC_T is exactly the same as in the original GPC.

Chapter 10
EDF in Real-Time Calculus

Response time analysis (RTA) is one of the key problems in real-time system design. This chapter proposes new RTA methods for EDF scheduling, with general system models where workload and resource availability are represented by request/demand bound functions and supply bound functions. The main idea is to derive response time upper bounds by lower-bounding the slack times. We first present a simple over-approximate RTA method, which lower bounds the slack time by measuring the "horizontal distance" between the demand bound function and the supply bound function. Then we present an exact RTA method based on the above idea but eliminating the pessimism in the first analysis. This new exact RTA method not only allows to precisely analyze more general system models than existing EDF RTA techniques, but also significantly improves analysis efficiency. Experiments are conducted to show efficiency improvement of our new RTA technique, and tradeoffs between the analysis precision and efficiency of the two proposed methods are discussed. We also illustrate the application of the proposed RTA techniques to Real-Time Calculus for the analysis of components with EDF scheduling in a distributed computing environment.

10.1 Introduction

RTA is one of the most important problems in real-time system design. RTA is not only useful to perform local schedulability test on single processors, but also plays important roles in the analysis of more complex real-time systems, e.g., distributed systems where the completion of a task generates outputs triggering computation or communication tasks on subsequent infrastructures [41, 75, 194, 201]. Since the completion time of the preceding task decides the release times of subsequent tasks, one can use RTA to bound the completion time of each task and decide the "release jitter" of the subsequent tasks.

© Springer International Publishing Switzerland 2016
N. Guan, *Techniques for Building Timing-Predictable Embedded Systems*,
DOI 10.1007/978-3-319-27198-9_10

EDF is a widely used real-time scheduling algorithm. Spuri [94] developed EDF RTA techniques for periodic tasks with extensions of jitters and sporadic bursts. It turns out that the RTA problem of EDF is more difficult than that of fixed-priority scheduling. Although RTA for fixed-priority scheduling has been extended to general workload and resource models represented by request bound functions (arrival curves) and supply bound functions (service curves) [202], no such work has been done for EDF to our best knowledge.

Spuri's EDF RTA technique relies on enumerating a (typically very large) number of concrete release patterns as the candidates of the worst-case scenario. On one hand, this requires high computational effort. On the other hand, this complicates its extension to more expressive models, as the technique needs to be customized to include new features of each new model. This is particularly hindering when workload and resource is specified in a rather abstract way.

In this chapter we present EDF RTA methods for general models with workload and resource represented by request/demand bound functions and supply bound functions. These general models are used in design and analysis frameworks such as Real-Time Calculus [203] and SymTA/S [75]. The key insight is that in EDF it is easier to derive response time upper bounds indirectly by lower-bounding the slack times. This not only allows us to perform RTA with general workload and resource models, but also can greatly improve the analysis efficiency.

More specifically, we first present a simple over-approximate RTA method, which lower bounds the slack times by measuring the "horizontal distance" between the demand bound function and the supply bound function. Then we present an *exact* RTA based on the similar idea but eliminating the pessimism in the first analysis. Experiments show that the new RTA technique can greatly improve the analysis efficiency comparing with Spuri's RTA. We also discuss the tradeoff between the analysis precision and efficiency of the two methods proposed in this chapter.

10.2   Preliminaries

10.2.1   Resource Model

We consider a processing platform with capacity characterized by a *supply bound function* $\mathsf{sbf}(\delta)$ [204, 205], which quantifies the minimal cumulative computation capacity provided by the processing platform within a time interval of length δ. The supply bound function is essentially the same as the lower service curve in Real-Time Calculus [203].

sbf is a continuous function. The pseudo-inverse function of sbf, denoted by $\overline{\mathsf{sbf}}$, characterizes the minimal interval length to provide a certain amount of computation capacity:

$$\overline{\mathsf{sbf}}(x) = \min\{\delta \,|\, \mathsf{sbf}(\delta) = x\}$$

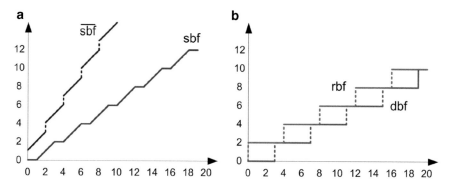

Fig. 10.1 Illustration of (**a**) sbf and $\overline{\text{sbf}}$, (**b**) rbf and dbf

Figure 10.1a illustrates sbf and its pseudo-inverse $\overline{\text{sbf}}$ of a TDMA resource with a period of 4 and slot size of 3.

10.2.2 Workload Model

The task system τ consists of $\mathcal{N}$ independent tasks $\{\tau_1, \tau_2, \cdots, \tau_{\mathcal{N}}\}$. Each task τ_i releases infinitely many *jobs*. We use J_i^j to denote the jth job released by task τ_i. For simplicity, we also omit the superscript and use J_i to denote a job released by τ_i when the context is clear.

Each task τ_i has a relative deadline D_i, which specifies the requirement that the computation demand of each job must be finished no later than D_i time units after its release time.

The workload of a task τ_i is characterized by a *request bound function* $\text{rbf}_i(\delta)$ [43], which quantifies the maximum cumulative execution requests that could be generated by jobs of τ_i released within a time interval of length δ.

rbf_i is a staircase function, and it is essentially the same as the upper arrival curve in Real-Time Calculus [203] and SymTA/S [75]. As a common assumption in EDF scheduling analysis [30, 94], there exists a bounded number $\mathcal{L}'$ such that

$$\text{rbf}(\mathcal{L}') \leq \text{sbf}(\mathcal{L}')$$

which guarantees that in long term the system source supply is no smaller than the total execution demand. We let $\mathcal{L} = \mathcal{L}' + d$ where d is the largest relative deadline among all tasks.

In the analysis of EDF scheduling, an important concept is the *demand bound function* dbf_i [30], which can be obtained by horizontally "shifting" rbf_i rightwards for D_i time units:

$$\text{dbf}_i(\delta) = \begin{cases} 0 & \delta < D_i \\ \text{rbf}_i(\delta - D_i) & \delta \geq D_i \end{cases} \tag{10.1}$$

Figure 10.1b illustrates the rbf and dbf of a sporadic task [30] with period of 4, relative deadline of 3, and worst-case execution time of 2. We define $\mathsf{sbf}(\delta)$ and $\mathsf{dbf}(\delta)$ as the total demand bound function of the system:

$$\mathsf{rbf}(\delta) = \sum_{\forall \tau_i \in \tau} \mathsf{rbf}_i(\delta) \quad \text{and} \quad \mathsf{dbf}(\delta) = \sum_{\forall \tau_i \in \tau} \mathsf{dbf}_i(\delta)$$

Finally, we assume the number of points where rbf and dbf "steps" in a unit-length interval is bounded by a constant, to exclude the case that rbf/dbf curves are "infinitely complex."

10.2.3   EDF Scheduling and Worst-Case Response Time

When a job of task τ_i is released at time t, its *absolute deadline* is at $t + D_i$. The task system is scheduled by EDF algorithm, which assigns priorities to *active* jobs (jobs that have been released but not finished yet) according to their absolute deadlines: the earlier deadline and higher priority. In case of multiple active jobs having the same absolute deadline, the EDF scheduler may prioritize *any* of them for execution. The aim of this chapter is to calculate the worst-case response time R_i of each task:

Definition 10.1 (Worst-Case Response Time). The worst-case response time R_i of a task τ_i is the length of the *longest* interval from a job's release till its completion.

Note that we allow the case of $R_i > D_i$. A job finishes after its absolute deadline is called a *tardy* job [150]. Although a tardy job cannot finish computation before the expected deadline, it is still interesting to know its *tardiness*, i.e., how much it lags behind, in many soft real-time systems [150, 206].

10.2.4   Review of Spuri's RTA for EDF

Spuri [94] introduced an RTA method for sporadic tasks with jitters and sporadic bursts on a fully dedicated processor ($\mathsf{sbf}(\delta) = \delta$). For simplicity of presentation, we review Spuri's RTA technique with periodic tasks. Details about handling jitters and sporadic bursts can be found in [94].

Each task τ_i is characterized by three parameters: worst-case execution time C_i, relative deadline D_i, and period T_i. The worst-case response time R_i of τ_i is calculated by

$$R_i = \max_{\forall p \in P, a \in A_p,} \mathscr{R}_i(a, p)$$

where

$$P = [1, \lceil \mathscr{L}/T_i \rceil]$$

$$A_p = \{x - (p-1)T_i - D_i | x \in X \cap [(p-1)T_i + D_i, pT_i + D_i]\}$$

$$X = \bigcup_{\forall \tau_j \in \tau, q \in [1, \lceil \frac{\mathscr{L}}{T_j} \rceil]} \{(q-1)T_j + D_j\}$$

$$\mathscr{R}_i(a, p) = w(a, p) - a - (p-1)T_i$$

and $w(a, p)$ is the minimal solution of equation

$$w(a, p) = pC_i + \sum_{\tau_j \neq \tau_i} W_j(w(a, p), \ a + (p-1)T_i + D_i)$$

where $W_j(x, y) = \min(\lceil \frac{x}{T_j} \rceil, \lfloor \frac{y - D_j}{T_j} \rfloor + 1) \cdot C_j$.

Spuri's analysis is rather complicated, even with the simple sporadic task and fully dedicated resource model. Extending it to more expressive models could be difficult and error-prone. On the other hand, Spuri's analysis contains tremendous redundant computation, which leads to low analysis efficiency. The overall complexity of Spuri's RTA to a periodic task set is $O(\mathcal{N}\mathcal{T}\mathscr{L}'^2)$, where $\mathcal{N}$ is the total number of tasks, $\mathcal{T}$ is the maximal period among all tasks, and $\mathscr{L}'$ is the maximal busy period size. The target of this chapter is to overcome the above problems, by providing EDF RTA techniques that are more general, more efficient, and easier to understand and remember.

10.3 Over-Approximate RTA

In this section, we first introduce a simple *over-approximate* RTA method. After presenting the analysis, we also use an example to explain why it may over-estimate the response time. Then in Sect. 10.4, we will reuse these insights and present our second RTA method which yields *exact* results.

Unlike traditional RTA techniques [31, 35, 94], our RTA method bounds the the response times *indirectly*: it calculates a *lower* bound on the *worst-case slack time* of the task, by which the response time upper bound can be easily obtained.

Definition 10.2 (Worst-Case Slack Time). The worst-case slack time S_i of a task τ_i is the length of the *shortest* interval from a job's completion till its absolute deadline.

The task's worst-case response time is calculated by

$$R_i = D_i - S_i$$

Note that R_i is greater than D_i when S_i is negative.

We can safely lower bound the slack time of each task by

Theorem 10.1. *The slack time of task τ_i is bounded from below by*

$$S_i^* = \min_{\forall \delta: \mathscr{L} \geq \delta \geq D_i} \left\{ \delta - \overline{\mathsf{sbf}}(\mathsf{dbf}(\delta)) \right\} \qquad (10.2)$$

Proof. We prove the theorem by contradiction. Suppose at run-time task τ_i has a job J_i whose slack time, denoted by S', is strictly smaller than S_i^*. Let t_r, t_d, and t_f be the release time, absolute deadline, and finish time of J_i, respectively. Let t_o be the earliest time instance before t_f such that at any time instant in $[t_o, t_f]$ there is at least one active job with deadline no later than t_d. Let $\ell = t_d - t_o$.

The total amount of workload (of jobs with deadline no later than t_d) in $[t_o, t_d]$ is bounded by $\mathsf{dbf}(\ell)$, and it takes the resource for at most $\overline{\mathsf{sbf}}(\mathsf{dbf}(\ell))$ time units to provide enough capacity to finish it. So we know there exists a time point in $[t_o, t_o + \overline{\mathsf{sbf}}(\mathsf{dbf}(\ell))]$ at which the processor is idle or executing jobs with deadline later than t_d. By the definition of t_o we know that this time point is not in $[t_o, t_f)$, so

$$t_o + \overline{\mathsf{sbf}}(\mathsf{dbf}(\ell)) \geq t_f \qquad (10.3)$$

Since J_i itself is an active job at t_r, by the definition of t_o we know $t_o \leq t_r$ and thus $\ell \geq D_i$. On the other hand, the length of $(t_o, t_r]$ is bounded by $\mathscr{L}'$ and the length of $(t_r, t_d]$ is bounded by d, so $\ell \leq \mathscr{L}$. In summary, $\mathscr{L} \geq \ell \geq D_i$. Then by (10.2) we have $S_i^* \leq \ell - \overline{\mathsf{sbf}}(\mathsf{dbf}(\ell))$, and by $S' < S_i^*$ we have $S' < \ell - \overline{\mathsf{sbf}}(\mathsf{dbf}(\ell))$, i.e., $t_d - S' > t_d - \ell + \overline{\mathsf{sbf}}(\mathsf{dbf}(\ell))$. Then we apply substitutes $t_o = t_d - \ell$ and $t_f = t_d - S'$ to get $t_f > t_o + \overline{\mathsf{sbf}}(\mathsf{dbf}(\ell))$, which contradicts (10.3). $\qquad \square$

Intuitively, the slack time lower bound stated in the above theorem is the minimal "horizontal distance" between dbf and sbf (in the range of $\delta \geq D_i$). Note that if $\mathsf{dbf}(\delta)$ is larger than $\mathsf{sbf}(\delta)$ for some $\delta > D_i$, then S_i^* is negative.

Example 10.1. Suppose task set τ consists of three sporadic tasks τ_1, τ_2, and τ_3 with parameters as follows: $\{C_1 = 1, \ T_1 = D_1 = 4\}$, $\{C_2 = 1, \ T_2 = D_2 = 12\}$, and $\{C_3 = 3, \ T_3 = D_3 = 16\}$. The task set is scheduled by EDF on a TDMA resource with period of 4 and slot size of 3. By Theorem 10.1 we obtain the slack time lower bound of each tasks: $S_1^* = 2$ and $S_2^* = S_3^* = 4$, as illustrated in Fig. 3.2a. So the worst-case response time of τ_1, τ_2, and τ_3 is bounded by $D_1 - 2 = 2$, $D_2 - 4 = 8$, and $D_3 - 4 = 12$, respectively.

The slack time lower bound S_i^* is safe, but in general pessimistic, as shown in the following example.

Fig. 10.2 Illustration of
Examples 10.1 and 10.2.
(**a**) The computation of S_i^*.
(**b**) The pessimism of S_i^*

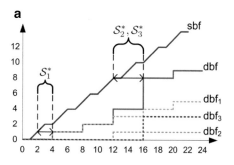

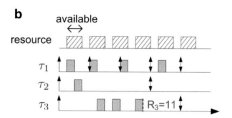

Example 10.2. When we analyze τ_3 in the above example, the minimal horizontal distance between $\mathsf{sbf}(\delta)$ and $\mathsf{dbf}(\delta)$ (only considering the part on $\overline{\mathsf{dbf}(\delta)}$ with $\delta \geq D_3$) occurred at $\delta = 16$, with $\mathsf{dbf}(16) = 8$ and $\overline{\mathsf{sbf}(\mathsf{dbf}(16))} = \overline{\mathsf{sbf}(8)} = 12$. The slack bound of τ_3 is $12 - 8 = 4$, and its response time bound is $D_3 - 4 = 12$. However, if we simulate the worst-case workload in a time interval of length 16 as in Fig. 10.2b, the response time is actually 11 (which is indeed the worst-case response time of τ_3). This is because, although the workload of the last job of τ_1 is included in $\mathsf{dbf}(16)$, this job is actually released after the finish time of the analyzed job of τ_3 and thus does not really contribute to the interference.

10.3.1 Special Case Where S_i^* Is Exact

Although S_i^* is generally a pessimistic bound, it is indeed the exact answer in the following special case:

Theorem 10.2. S_i^* *is the exact worst-case slack time of task τ_i if $S_i^* < 0$.*

Due to space limit we omit the formal proof but only provide the intuition: According to the standard dbf-based EDF schedulability test [30], a task τ_i is not schedulable iff sbf and rbf cross each other at some point no smaller than D_i, i.e., $S_i^* < 0$. When a job J_i finishes after its deadline, the jobs with release time and deadline in $[t_o, t_d]$ all can interfere with J_i. So the over-estimation problem in the above example does not exist, and $\mathsf{dbf}(t_d - t_o)$ precisely quantifies the total workload that needs to be finished before J_i is done.

When S_i^* is negative, τ_i is a *tardy* task. Theorem 10.2 says that $|S_i^*|$ is the exact worst-case *tardiness* of a tardy task τ_i.

10.3.2 Algorithmic Implementation and Complexity

The computation of S_i^* for different tasks is essentially the same. The only difference is that for each task τ_i we only need to visit δ no smaller than its D_i. So we only need to "scan" the curve $\mathsf{dbf}(\delta)$ for one time, to compute the slack time bounds of *all* tasks in τ.

Algorithm 6 shows the pseudo-code of the algorithm to compute $S_1^*, \cdots, S_{\mathcal{N}}^*$. For simplicity of presentation, we assume all tasks have different relative deadlines, and tasks are sorted in increasing order of their relative deadlines. The case where multiple tasks have the same relative deadline can be easily handled with minor revisions of the presented algorithm. Intuitively, the algorithm first calculates the minimal "horizontal distance" between dbf and sbf with δ values in each segment $[D_i, D_{i+1})$, recorded by s_i. With these s_i, the slack time bound of all tasks can be calculated in $O(\mathcal{N})$ time.

Since sbf is continuous and dbf is a staircase function, the candidate values of δ are the points in $[D_1, \mathcal{L}]$ at which $\mathsf{dbf}(\delta)$ is not differentiable i.e., where dbf "steps." So the number of candidate values of δ is bounded by $O(\mathcal{L})$ (recall that the number of points where dbf "steps" in a unit length is bounded by a constant). Moreover, the second for-loop iterates for $\mathcal{N}$ times. So the overall complexity of Algorithm 6 is $O(\mathcal{L} + \mathcal{N})$.

```
 1: s₁ = s₂ = ··· = s_𝒩 = +∞
 2: i = 1
 3: for each candidate value of δ do
 4:     if δ ≥ D_{i+1} then
 5:         i ← i + 1
 6:     end if
 7:     sᵢ ← min(sᵢ, δ − sbf̄(dbf(δ)))
 8: end for
 9: for i ← 𝒩 ··· 1 do
10:     Sᵢ* ← sᵢ
11:     s_{i−1} ← min(sᵢ, s_{i−1}) // does not execute when i = 1
12: end for
13: return S₁*, S₂*, ··· , S_𝒩*
```

Algorithm 6: Pseudo-code of the algorithm implementing Theorem 10.1.

10.4   Exact RTA

S_i^* is a pessimistic slack time lower bound since it ignores the fact that the workload released after the finish time of the analyzed job should not be included into the interference calculation. In this section we address this problem and present an exact RTA method.

As presented in last section, S_i^* is the exact worst-case slack time if τ_i is a tardy task (S_i^* is negative). So in this section we focus on the case that τ_i is *not* tardy, i.e., it holds

$$\forall \delta \geq D_i : \mathsf{dbf}(\delta) \leq \mathsf{sbf}(\delta) \tag{10.4}$$

We first define a task's *mixed bound function*:

Definition 10.3 (Mixed Bound Function). For any $\delta \geq \gamma \geq 0$, the *mixed bound function* of τ_i is defined as

$$\mathsf{mbf}_i(\delta, \gamma) = \min(\mathsf{dbf}_i(\delta), \mathsf{rbf}_i(\gamma))$$

Figure 10.3 illustrates $\mathsf{mbf}_i(\delta, \gamma)$. We consider two time intervals $[t_o, t_o + \delta]$ and $[t_o, t_o + \gamma]$ which both start at time t_o. $\mathsf{mbf}_i(\delta, \gamma)$ captures the workload of τ_i's jobs that are released in $[t_o, t_o + \gamma]$ and with deadline no later than $t_o + \delta$. In (a), while $\mathsf{dbf}(\delta)$ includes all of the five jobs, $\mathsf{sbf}(\gamma)$ only includes the first four jobs since the last job is released after $t_o + \gamma$, so $\mathsf{mbf}_i(\delta, \gamma)$ equals the total workload of the first four jobs. In (b), while all the three jobs are released in $[t_o, t_o + \gamma]$, $\mathsf{dbf}(\delta)$ excludes the last job as its deadline is later than $t_o + \delta$.

We define the total mixed bound function of the system:

$$\mathsf{mbf}(\delta, \gamma) = \sum_{\forall \tau_i \in \tau} \mathsf{mbf}_i(\delta, \gamma)$$

Note that $\mathsf{mbf}(\delta, \gamma)$ does *not* necessarily equal $\min(\mathsf{dbf}(\delta), \mathsf{sbf}(\gamma))$. With this new function $\mathsf{mbf}(\delta, \gamma)$, we can compute a task's *exact* worst-case slack time:

Theorem 10.3. *The exact worst-case slack time of task τ_i is computed by*

$$S_i = \min_{\forall \delta : \mathscr{L} \geq \delta \geq D_i} \max_{\forall \gamma : \gamma \leq \delta} \{\delta - \gamma \,|\, \mathsf{mbf}(\delta, \gamma) \leq \mathsf{sbf}(\gamma)\} \tag{10.5}$$

Fig. 10.3 Illustration of the mixed bound function $\mathsf{mbf}_i(\delta, \gamma)$

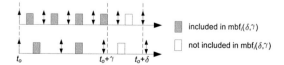

Proof. First, we show S_i is well defined in the sense that

$$\{(\delta, \gamma) | \mathscr{L} \geq \delta \geq D_i \wedge \gamma \leq \delta \wedge \mathsf{mbf}(\delta, \gamma) \leq \mathsf{sbf}(\gamma)\} \qquad (10.6)$$

is *not* empty and thus (10.5) always returns an answer.

Let x be a number in $[D_i, \mathscr{L}]$, then by (10.4) we know

$$\mathsf{dbf}(x) \leq \mathsf{sbf}(x) \Rightarrow \mathsf{mbf}(x, x) \leq \mathsf{sbf}(x)$$

so any (δ, γ) s.t. $\mathscr{L} \geq \delta \geq D_i \wedge \delta = \gamma$ is contained in (10.6).

In the following we prove S_i is both a *safe* and *tight* lower bound of τ_i's slack time.

Safety: We shall prove that the slack time of any job released by τ_i is no smaller than S_i. We prove it by contradiction. Suppose J_i is a job of task τ_i with slack time $S' < S_i$. Let t_r, t_d, and t_f be the release time, absolute deadline, and finish time of J_i, respectively. Let t_o be the earliest time instance before t_f such that at any time instant in $[t_o, t_f]$ there is at least one active job with deadline no later than t_d. Let $\delta' = t_d - t_o$. Since J_i is active at t_r, we know $t_o \leq t_r$ and thus $\delta' \geq D_i$. The length of $[t_o, t_r]$ is bounded by $\mathscr{L}'$ and the length of $[t_r, t_d]$ is bounded by d, so $\delta' \leq \mathscr{L} = \mathscr{L}' + d$. In summary, $\mathscr{L} \geq \delta' \geq D_i$. Then by the definition of S_i and $S' < S_i$ we have

$$S' < \max_{\gamma \leq \delta'} \{\delta' - \gamma \mid \mathsf{mbf}(\delta', \gamma) \leq \mathsf{sbf}(\gamma)\}$$

Let γ' be the *smallest* assignment of γ s.t. $\mathsf{mbf}(\delta', \gamma) \leq \mathsf{sbf}(\gamma)$ (thus $\delta' - \gamma$ is maximized), then we have

$$S' < \delta' - \gamma' \Rightarrow t_d - S' > t_d - \delta' + \gamma' \Rightarrow t_f > t_o + \gamma' \qquad (10.7)$$

On the other hand, since $\mathsf{mbf}(\delta', \gamma') \leq \mathsf{sbf}(\gamma')$, the workload of all jobs released in $[t_o, t_o + \gamma']$ and with deadline no later than $t_o + \delta' = t_d$ has been finished by $t_o + \gamma'$, and particularly, J_i has been finished by $t_o + \gamma'$, which contradicts (10.7).

Tightness: We shall construct a scenario where the slack time of a job of τ_i is exactly S_i. Let δ' and γ' be the assignments of δ and γ that give the value of S_i in (10.5). Let each task release its first job at time t_o and release as much workload as possible since then (following its rbf_i). Moreover, we move the release time of the last job of the analyzed task τ_i released in $[t_o, t_o + \delta' - D_i]$, denoted by J_i, such that its deadline aligns with $t_o + \delta'$. Note that $[t_o, t_o + \delta' - D_i]$ is well defined since $\delta' \geq D_i$. Let J_i have the lowest priority among all the jobs with deadlines at $t_o + \delta'$. Since γ' is the assignment of γ maximizing $\delta' - \gamma$ with this particular δ', we know:

$$\forall \gamma'' < \gamma' : \mathsf{mbf}(\delta', \gamma'') > \mathsf{sbf}(\gamma'') \qquad (10.8)$$

Under the particular release pattern described above, $\mathsf{mbf}(\delta', \gamma'')$ is the exact total workload of jobs (of all tasks) released in $[t_o, t_o + \gamma'')$ and having priority no lower than J_i. By (10.8) we know $t_o + \gamma'$ is the first time point after t_o at which the processor is idle or executing jobs with priority lower than J_i. Therefore J_i is finished exactly at $t_o + \gamma'$ and its slack time equals $\delta' - \gamma'$, i.e., equals S_i.

By examining (10.5) we can get a general property of EDF scheduling concerning the relation of tasks' relative deadlines and worst-case slack times:

Corollary 10.1. *For any two tasks τ_i and τ_j in a task set τ scheduled by EDF, it holds*

$$D_i = D_j \Rightarrow S_i = S_j \quad and \quad D_i > D_j \Rightarrow S_i \geq S_j$$

10.4.1 Algorithmic Implementation and Complexity

A naive way to compute S_i using (10.5) is enumerating all the combinations of δ and γ candidate values satisfying $\mathscr{L} \geq \delta \geq D_i$ and $\delta \geq \gamma$. This could be inefficient when $\mathscr{L}$ is large. Algorithm 7 presents the pseudo-code for a more efficient implementation.

Similar to Algorithm 6, Algorithm 7 also integrates the slack time bounds computation of *all* tasks, using s_i to keep track of the local minimum in each segment $[D_i, D_{i+1})$. Algorithms 7 and 6 differ in the first *for-loop*. Two optimizations are applied to speed up the analysis. Firstly, to compute

```
1:  s₁ = s₂ = · · · = s_𝒩 = +∞
2:  i = 1
3:  for each candidate value of δ (in increasing order) do
4:      if δ ≥ D_{i+1} then
5:          i ← i + 1
6:      end if
7:      if δ − sbf(dbf(δ)) < s_i then
8:          γ_old ← 0
9:          γ_new ← sbf(mbf(δ, 0))
10:         while γ_new ≠ γ_old do
11:             γ_old ← γ_new
12:             γ_new ← sbf(mbf(δ, γ_old))
13:         end while
14:         s_i ← min(s_i, δ − γ_new)
15:     end if
16: end for
17: for i = 𝒩 · · · 1 do
18:     S_i ← s_i
19:     s_{i−1} ← min(s_i, s_{i−1})  // does not execute when i = 1
20: end for
21: return S₁, S₂, · · · , S_𝒩
```

Algorithm 7: Pseudo-code of the algorithm implementing Theorem 10.3.

$$\max_{\forall \gamma: \gamma \leq \delta} \{\delta - \gamma \,|\, \mathsf{mbf}(\delta, \gamma) \leq \mathsf{sbf}(\gamma)\}$$

with a particular δ, we use the well-known iterative fixed-point calculation technique [31] instead of searching over all possible γ values (lines 8 to 14).

Secondly, for each candidate value of δ, we first check whether $\delta - \overline{\mathsf{sbf}}(\mathsf{dbf}(\delta))$ is smaller than the current s_i. If not, then we can safely skip further calculation regarding this δ value. This is because $\delta - \overline{\mathsf{sbf}}(\mathsf{dbf}(\delta))$ is a lower bound of the slack time with this particular δ. If this lower bound is already greater than s_i, then using $\mathsf{mbf}(\delta, \gamma)$ to further refine this bound (make it potentially bigger) will not yield a result smaller than s_i. This optimization is very effective in practice. Typically, the worst-case slack time of a task occurs with relatively small δ values. Since we check the δ candidates in increasing order, we can obtain small slack time estimations with smaller δ values, and skip the fixed-point computation regarding γ with large δ values.

Similar to Algorithm 6, the candidate values of δ are the points in $[D_1, \mathscr{L}]$ where dbf "steps." The overall complexity of Algorithm 7 is $O(\mathscr{L}^2 + \mathscr{N})$.

10.5 Evaluation

10.5.1 Efficiency Improvement

We first evaluate the analysis efficiency improvement of the exact slack (response) time analysis in Sect. 10.4 comparing with Spuri's method in [94]. We adopt the sporadic task model [30] and a fully dedicated processor ($\mathsf{sbf}(\delta) = \delta$), with which Spuri's method is also applicable. Tasks are randomly generated, with the periods (T_i) uniformly distributed in $[100, 1000]$, utilizations (C_i/T_i) uniformly distributed in $[0.01, 0.2]$, and the ratio between the relative deadline and period (D_i/T_i) uniformly distributed in $[0.8, 1]$. We generate task sets as follows: A task set of 2 tasks is generated and analyzed by both our exact RTA and Spuri's RTA. Then we increase the number of tasks by 1 to generate a new task set. This process is iterated until the total utilization exceeds 100 %. The whole procedure is then repeated, starting with a new task set of 2 tasks, until a reasonably large sample space has been generated.

Figure 10.4 shows the *speedup* of our new exact RTA over Sprui's. For each task set, the *speedup* is the ratio between the analysis time by Sprui's RTA and that by our exact RTA. A point in the curve represents the average *speedup* of all task sets generated in a certain scope of system utilization corresponding to the abscissa.

From Fig. 10.4 we can see that the analysis efficiency improvement of our new RTA over Spuri's RTA is significant, especially with task sets of high total utilizations. The reason is twofold. First, Sprui's RTA analyzes each task separately, while our RTA integrates the analysis of all tasks and only "scans" the demand bound function curve once. A task set with higher utilization typically contains more tasks, so the efficiency improvement of our RTA is greater. Second, when the task

Fig. 10.4 Speedup of our
exact RTA comparing with
Spuri's method

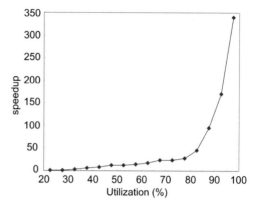

Fig. 10.5 Comparison of the
two RTA methods

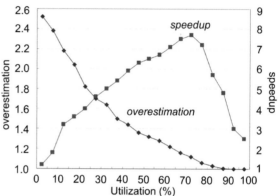

set's total utilization is close to 100%, $\mathscr{L}$ is typically very large. The efficiency of
Sprui's RTA is sensitive to $\mathscr{L}$: the number of different combinations to be checked
by Sprui's RTA grows rapidly as $\mathscr{L}$ increases. However, thanks to the optimization
of skipping the refinement with δ values satisfying $\delta - \overline{\mathsf{sbf}}(\mathsf{dbf}(\delta)) < s_i$ (see the last
second paragraph of Sect. 10.4), our exact RTA can skip the computation between
lines 8 and 14 of Algorithm 7 for most large δ values, and thus is less sensitive to
the growth of $\mathscr{L}$.

10.5.2 Comparing the Two Proposed Methods

In the following we compare the two RTA methods of this work, regarding their
efficiency and precision. We use the same strategy as above to generate task sets. In
Fig. 10.5, each point on curve *over-estimation* represents the average ratio between
the response time bounds obtained by the approximate RTA and by the exact
RTA for task sets in a certain total utilization scope. Each point on curve *speedup*
represents the average ratio between the analysis time of the exact RTA and the
approximate RTA in a certain total utilization scope.

By curve *over-estimation* we see that the precision of the approximate RTA increases as the total utilization increases. This is because that in task sets with higher total utilization, tasks' response times are typically closer to relative deadlines, so it has lower chance for the approximate RTA to mis-include the workload released after the analyzed job's finish time.

By curve *speedup* we see that the efficiency gap between the exact and the approximate RTA methods is smaller for task sets with either very low or very high total utilization. When the task set has very low total utilization, the iterative fixed-point calculation regarding γ typically converges very quickly, so the analysis time by the two methods is close. When the task set has very high total utilization, $\mathscr{L}$ is typically very large. As we discussed in last subsection, the exact RTA skips the fixed-point calculation regarding γ for a large portion of δ values. In other words, for most δ values the computation effort of both methods is the same, so the gap between their overall analysis time is small.

From the above results we can see that as the total utilization increases, the approximate RTA becomes more precise and gets more rewards in efficiency. For tasks with very high total utilization ($\geq 80\%$), the approximate RTA is almost as precise as the exact RTA, but at the same time the efficiency of the exact RTA is also catching up. It's up to the system designer to choose the proper analysis method according to their efficiency and precision requirements. But at least one can draw a clear conclusion that for task systems with very low total utilization, it does not make much sense to use the approximate RTA as it leads to rather imprecise results but benefits little in efficiency.

10.6   Analysis of EDF Component in RTC

An important application of RTA for EDF scheduling is the analysis of distributed systems, where the processed tasks generate output events triggering tasks on the subsequent computation units or communication interconnections.

Real-Time Calculus [203] is a general framework for performance analysis of distributed embedded systems, which is rooted in and significantly extending Network Calculus theory [187]. In RTC, workload is represented by a pair of upper and lower arrival curves α^u, α^l and resource availability is represented by a pair of upper and lower service curves β^u, β^l. As mentioned in Sect. 10.2, essentially α^u (defined on the basis of workload rather number of events) is the same as the request bound function rbf, and β^l is the same as the lower bound function sbf. The RTC framework connects multiple components into a network to model systems with resource sharing and networked structures. At each component, RTC generates the output arrival and service curves from the input arrival and service curves, and then propagates the output curves to the subsequent components. So one of the most important problems in RTC is how to compute the output curves from input curves for various types of components modeling different resource arbitration mechanisms.

The state-of-the-art techniques for analyzing EDF scheduling in RTC was introduced by Perathoner [207], where the output curves and EDF component are computed by

$$\alpha_i'^u(\Delta) = \min(\alpha_i^u(\Delta + D_i - BCET_i), \tilde{\alpha}_i'^u(\Delta)) \qquad (10.9)$$

$$\alpha_i'^l(\Delta) = \max(\alpha_i^l(\Delta + D_i - BCET_i), \tilde{\alpha}_i'^l(\Delta)) \qquad (10.10)$$

where $BCET_i$ is the best-case processing time of a single request. A detailed explanation of $\tilde{\alpha}_i'^u(\Delta)$ and $\tilde{\alpha}_i'^l(\Delta)$ can be found in [207]. In the following we focus on the first items in the min and max operation. Intuitively, $\alpha_i^u(\Delta + (D_i - BCET_i))$ and $\alpha_i^l(\Delta - (D_i - BCET_i))$ bounds the output curves by counting the delay variance of each request: the maximal delay is bounded by its relative deadline D_i and the minimal delay is its best-case processing time.

Since the worst-case response time R_i also bounds the maximal delay of each request, we can use R_i to replace D_i in (10.9) and (10.10) to compute more precise output curves. Figure 10.6 shows an example of RTC analysis network of three workload streams flowing through three EDF components. The initial input arrival curves model the PJD tasks [40], and the input service curves model the TDMA resources, with the parameters shown in Tables 10.1 and 10.2. Figure 10.7 depicts the output arrival curves at each component. In each picture, the red curves are generated using the current implementation of EDF components in RTC Toolbox [80] using (10.9) and (10.10), and the blue curves are generated using the exact response time bound R_i derived by the method of Sect. 10.4 to replace the deadline D_i in (10.9) and (10.10). In Fig. 10.7f the red curves and the blue curves are exactly the same and only the blue ones are displayed. The end-to-end delay bounds (sum of delay bounds of each component) of the three streams using the original method and our new techniques are shown in Table 10.3. From these results we can see that our RTA can help to greatly improve the analysis precision of EDF components in RTC.

Apart from the imprecise computation of output curves, the original method by Perathoner also limits the modeling power of RTC. The deadline of each request in EDF can be viewed as the metric to decide its priority, i.e., the so-called *priority point*. The priority point is not necessarily the same as the deadline, but can be any instant in time. For example, if the priority point of each request aligns

Fig. 10.6 An example of RTC analysis network with EDF components

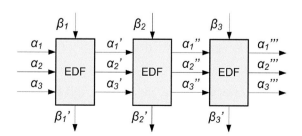

Table 10.1 PJD-task parameters of α_1, α_2, and α_3

Stream no.	Period	Jitter	Distance	Deadline
1	12	20	6	12
2	20	4	18	16
3	22	10	20	18

Table 10.2 TDMA parameters of β_1, β_2, and β_3

Resource no.	Slot size	Period
1	8	10
2	10	12
3	1	1

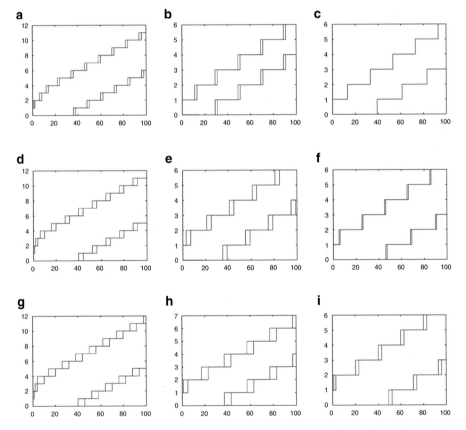

Fig. 10.7 Illustration of the output arrival curves in the example, (**a**) α'_1, (**b**) α'_2, (**c**) α'_3, (**d**) α''_1, (**e**) α''_2, (**f**) α''_3, (**g**) α'''_1, (**h**) α'''_2, (**i**) α'''_3

with its invocation time, the requests will be scheduled in a FIFO (first-in-first-out) manner. In general, the scheduling algorithms that define priority points for each task with a fixed offset with the release time of each job are called EDF-like scheduling algorithm. However, the current RTC framework interprets EDF

Table 10.3 The end-to-end delay bounds of each workload stream

Stream no.	Original method	Our new method
1	$22 = (8 + 8 + 6)$	$11 = (5+5+1)$
2	$25 = (9 + 10 + 6)$	$16 = (6+7+3)$
3	$26 = (9 + 10 + 7)$	$21 = (8 + 8 + 5)$

scheduling in a narrow sense and in general does not allow to model EDF-like scheduling algorithms except EDF and FIFO. Our new RTA technique not only improves the analysis precision, but also decouples the concept of deadline and priority point in EDF scheduling, and thereby supports modeling and analysis of a wide range of EDF-like scheduling policies in RTC.

10.7 Conclusions

RTA is not only useful in local schedulability test, but also lends itself to many complex design and analysis problems. Examples include distributed systems (as we discussed in Sect. 10.1) and the compositional scheduling problem [204, 205] (where the response time of a task on a certain level helps to tighten the resource supply bound to its inner task subsystem). The deadline adjustment technique with EDF [204, 208] can also be used to regulate task finish times in the above problems. However, when the scheduler is encapsulated but not open to designers for modification (which is common in component-based system design), the deadline adjustment technique is not applicable and one has to resort to RTA for finishing times characterization. We propose new RTA methods for EDF, with general system models represented by request/demand bound functions and supply bound functions. Our new RTA method not only allows to precisely analyze more general system models than existing EDF RTA techniques, but also significantly improves analysis efficiency.

References

1. ARTEMIS European Technology Platform, http://www.artemis-ju.eu/embedded_systems
2. V. Suhendra, T. Mitra, Exploring locking and partitioning for predictable shared caches on multi-cores, in *DAC*, 2008
3. A. Vajda, *Programming Many-Core Chips* (Springer, Dordrecht, 2011)
4. R. Wilhelm, J. Engblom, A. Ermedahl, N. Holsti, S. Thesing, D. Whalley, G. Bernat, C. FerdinanRd, R. Heckmann, T. Mitra, F. Mueller, I. Puaut, P. Puschner, J. Staschulat, P. Stenström, The worst-case execution-time problem overview of methods and survey of tools. ACM Trans. Embed. Comput. Syst. **7**(3), 36:1–36:53 (2008)
5. H. Theiling, Control flow graphs for real-time system analysis: reconstruction from binary executables and usage in ilp-based path analysis. Ph.D. thesis, Universitat des Saarlandes, 2002
6. C. Cullmann, F. Martin, Data-flow based detection of loop bounds, in *7th International Workshop on Worst-Case Execution Time (WCET) Analysis*, ed. by C. Rochange (Dagstuhl, Wadern). Internationales Begegnungs- und Forschungszentrum f"ur Informatik (IBFI) (Schloss Dagstuhl, Wadern, 2007)
7. J. Gustafsson, A. Ermedahl, C. Sandberg, B. Lisper, Automatic derivation of loop bounds and infeasible paths for wcet analysis using abstract execution, in *RTSS* (IEEE Computer Society, Washington, DC, 2006), pp. 57–66
8. C. Healy, M. Sjödin, V. Rustagi, D. Whalley, R. Van Engelen, Supporting timing analysis by automatic bounding of loop iterations. Real-Time Syst. **18**(2/3), 129–156 (2000)
9. F. Stappert, P. Altenbernd, Complete worst-case execution time analysis of straight-line hard real-time programs. J. Syst. Archit. **46**(4), 339–355 (2000)
10. P. Altenbernd, On the false path problem in hard real-time programs, in *Proceedings of the 8th Euromicro Workshop on Real-Time Systems*, 1996, pp. 102–107
11. C.A. Healy, D.B. Whalley, Automatic detection and exploitation of branch constraints for timing analysis. IEEE Trans. Softw. Eng. **28**(8), 763–781 (2002)
12. R. Wilhelm, P. Lucas, O. Parshin, L. Tan, B. Wachter, Improving the precision of WCET analysis by input constraints and model-derived flow constraints, in *Advances in Real-Time Systems*, ed. by S. Chakraborty, J. Eberspächer. Lecture Notes in Computer Science. (Springer, Berlin, 2011)
13. J. Engblom, Processor pipelines and static worst-case execution time analysis. Ph.D. Thesis, Uppsala University, 2002
14. X. Li, A. Roychoudhury, T. Mitra, Modeling out-of-order processors for software timing analysis, in *RTSS*, 2004, pp. 92–103

© Springer International Publishing Switzerland 2016
N. Guan, *Techniques for Building Timing-Predictable Embedded Systems*,
DOI 10.1007/978-3-319-27198-9

15. S. Thesing, Safe and precise wcet determination by abstract interpretation of pipeline models. Ph.D. thesis, Universitat des Saarlandes, 2004
16. S. Wilhelm, B. Wachter, Symbolic state traversal for wcet analysis, in *EMSOFT* (ACM, New York, NY, 2009), pp. 137–146
17. E. Althaus, S. Altmeyer, R. Naujoks, Precise and efficient parametric path analysis, in *LCTES* (ACM, New York, NY, 2011), pp. 141–150
18. Y.-T. Steven Li, S. Malik, Performance analysis of embedded software using implicit path enumeration, in *DAC* (ACM, New York, NY, 1995), pp. 456–461
19. H. Theiling, C. Ferdinand, R. Wilhelm, Fast and precise wcet prediction by separated cache and path analyses. Real-Time Syst. **18**(2/3), 157–179 (2000). doi:10.1023/A:1008141130870. http://dx.doi.org/10.1023/A:1008141130870
20. H. Theiling, Ilp-based interprocedural path analysis, in *EMSOFT '02* (Springer, London, 2002), pp. 349–363
21. D. Grund, J. Reineke, Abstract interpretation of fifo replacement, in *SAS* (Springer, Berlin/Heidelberg, 2009), pp. 120–136
22. N. Guan, M. Lv, W. Yi, G. Yu, Wcet analysis with mru caches: Challenging lru for predictability. in *RTAS*, 2012, pp. 55–64
23. D. Grund, J. Reineke, Toward precise plru cache analysis, in *WCET*, 2010, pp. 23–35
24. C. Ferdinand, Cache behavior prediction for real-time systems. Ph.D. thesis, Universitat des Saarlandes, 1997
25. F. Mueller, *Generalizing Timing Predictions to Set-Associative Caches*, vol. 0 (IEEE Computer Society, Los Alamitos, CA, 1997), p. 64
26. J. Reineke, D. Grund, C. Berg, R. Wilhelm, Timing predictability of cache replacement policies. Real-Time Syst. **37**(2), 99–122 (2007). doi:10.1007/s11241-007-9032-3. http://dx.doi.org/10.1007/s11241-007-9032-3
27. R. Wilhelm, D. Grund, J. Reineke, M. Schlickling, M. Pister, C. Ferdinand, Memory hierarchies, pipelines, and buses for future architectures in time-critical embedded systems. IEEE Trans. Comput. Aided Des. Integr. Circuits Syst. **28**(7), 966–978 (2009)
28. D. Grund, J. Reineke, Precise and efficient fifo-replacement analysis based on static phase detection, in *ECRTS* (IEEE Computer Society, Washington, DC, 2010), pp. 155–164
29. M. Lv, N. Guan, W. Yi, Q. Deng, G. Yu, Efficient instruction cache analysis with model checking, in *RTAS, Work-in-Progress Session*, 2010
30. S. Baruah, A. Mok, L. Rosier, Preemptively scheduling hard-real-time sporadic tasks on one processor, in *RTSS*, 1990
31. M. Joseph, P.K. Pandya, Finding response times in a real-time system. Comput. J. **29**(5), 390–395 (1986). doi:10.1093/comjnl/29.5.390. http://dx.doi.org/10.1093/comjnl/29.5.390
32. C.L. Liu, J.W. Layland, Scheduling algorithms for multiprogramming in a hard-real-time environment. J. ACM **20**(1), 46–61 (1973). doi:10.1145/321738.321743. http://doi.acm.org/10.1145/321738.321743
33. On-Line Applications Research Corporation (OAR), RTEMS Applications C User's Guide (2001)
34. J. Calandrino, H. Leontyev, A. Block, U. Devi, J. Anderson, Litmusrt: a testbed for empirically comparing real-time multiprocessor schedulers, in *RTSS*, 2006
35. J.P. Lehoczky, Fixed priority scheduling of periodic task sets with arbitrary deadlines, in *RTSS*, 1990
36. R.I. Davis, A. Burns, R.J. Bril, J.J. Lukkien, Controller Area Network (CAN) schedulability analysis: refuted, revisited and revised. Real-Time Syst. **35**(3), 239–272 (2007)
37. L. George, N. Rivierre, M. Spuri, Preemptive and non-preemptive real-time uni-processor scheduling. Technical Report, INRIA, 1996
38. L. Sha, R. Rajkumar, J. Lehoczky, Priority inheritance protocols: an approach to real-time synchronization. IEEE Trans. Comput. **39**(9), 1175–1185 (1990). doi:10.1109/12.57058. http://dx.doi.org/10.1109/12.57058

39. N. Audsley, A. Burns, M. Richardson, K. Tindell, A.J. Wellings, Applying new scheduling theory to static priority preemptive scheduling. Softw. Eng. J. **8**(5), 284–292 (1993). http://ieeexplore.ieee.org/xpl/articleDetails.jsp?arnumber=238595

40. A. Hamann, M. Jersak, K. Richter, R. Ernst, Design space exploration and system optimization with symta/s-symbolic timing analysis for systems, in *RTSS*, 2004

41. M. Gonzalez Harbour, J. Palencia Gutierrez, Schedulability analysis for tasks with static and dynamic offsets, in *RTSS*, 1998

42. K. Tindell, Adding time-offsets to schedulability analysis. Technical Report, 1994

43. S. Baruah, Dynamic- and static-priority scheduling of recurring real-time tasks. Real-Time Syst. **24**(1), 93–128 (2003). doi:10.1023/A:1021711220939. http://dx.doi.org/10.1023/A:1021711220939

44. M. Stigge, N. Guan, W. Yi, Refinement-based exact response time analysis. Technical Report, 2014

45. T.W. Kuo, A.K. Mok, Load adjustment in adaptive real-time systems, in *RTSS*, 1991

46. C. Lu, J.A. Stankovic, S.H. Son, G. Tao, Feedback control real-time scheduling: framework, modeling, and algorithms. Real-Time Syst. **23**(1–2), 85–126 (2002)

47. C.L. Liu, Scheduling algorithms for multiprocessors in a hard real-time environment, in *JPL Space Programs Summary*, 1969

48. J. Carpenter, S. Funk, P. Holman, A. Srinivasan, J. Anderson, S. Baruah, A categorization of real-time multiprocessor scheduling problems and algorithms, in *Handbook of Scheduling – Algorithms, Models, and Performance Analysis* (2004). http://www.crcnetbase.com/doi/abs/10.1201/9780203489802.ch30

49. R.I. Davis, A. Burns, A survey of hard real-time scheduling for multiprocessor systems. ACM Comput. Surv. **43**(4), 35:1–35:44 (2011)

50. B. Andersson, S. Baruah, J. Jonsson, Static-priority scheduling on multiprocessors, in *RTSS*, 2001

51. B. Andersson, Global static priority preemptive multiprocessor scheduling with utilization bound 38%, in *OPODIS*, 2008

52. T.P. Baker, Multiprocessor edf and deadline monotonic schedulability analysis, in *RTSS*, 2003

53. S.K. Baruah, Techniques for multiprocessor global schedulability analysis, in *RTSS*, 2007

54. M. Bertogna, M. Cirinei, G. Lipari, Improved schedulability analysis of edf on multiprocessor platforms, in *ECRTS*, 2005

55. N. Guan, M. Stigge, W. Yi, G. Yu, New response time bounds for fixed priority multiprocessor scheduling, in *Proceedings of the 30st IEEE Real-Time Systems Symposium (RTSS)*, 2009

56. B. Andersson, J. Jonsson, Some insights on fixed-priority preemptive non-partitioned multiprocessor scheduling. Technical Report, Chalmers University of Technology, 2001

57. S. Lauzac, R.G. Melhem, D. Mosse, Comparison of global and partitioning schemes for scheduling rate monotonic tasks on a multiprocessor, in *ECRTS*, 1998

58. S.K. Dhall, C.L. Liu, On a real-time scheduling problem. Oper. Res. **26**(1), 127–140 (1978)

59. B. Andersson, J. Jonsson, The utilization bounds of partitioned and pfair static-priority scheduling on multiprocessors are 50%, in *Euromicro Conference on Real-Time Systems (ECRTS)*, 2003

60. S. Baruah, N. Fisher, The partitioned multiprocessor scheduling of sporadic task systems, in *RTSS*, 2005

61. A. Burchard, J. Liebeherr, Y. Oh, S.H. Son, New strategies for assigning real-time tasks to multiprocessor systems. IEEE Trans. Comput. **44**(12), 1429–1442 (1995)

62. N. Fisher, S.K. Baruah, T.P. Baker, The partitioned scheduling of sporadic tasks according to static-priorities, in *ECRTS*, 2006, pp. 118–127

63. D. Oh, T.P. Baker, Utilization bounds for n-processor rate monotone scheduling with static processor assignment. Real-Time Syst. **15**(2), 183–192. doi:10.1023/A:1008098013753. http://dx.doi.org/10.1023/A:1008098013753 (1998)

64. S.K. Dhall, S. Davari, On a periodic real time task allocation problem, in *Annual International Conference on System Sciences*, 1986

65. E.G. Coffman Jr., M.R. Garey, D.S. Johnson, Approximation algorithms for bin packing: a survey, in *Approximation Algorithms for NP-Hard Problems*, ed. by D.S. Hochbaum (PWS Publishing Co., Boston, 1997), pp. 46–93. http://dl.acm.org/citation.cfm?id=241938.241940

66. J. Anderson, V. Bud, U.C. Devi, An EDF-based scheduling algorithm for multiprocessor soft real-time systems, in *ECRTS*, 2005

67. B. Andersson, K. Bletsas, S. Baruah, Scheduling arbitrary-deadline sporadic task systems multiprocessors, in *RTSS*, 2008

68. N. Guan, M. Stigge, W. Yi, G. Yu, Fixed-priority multiprocessor scheduling with Liu & Layland's utilization bound, in *RTAS*, 2010

69. S. Kato, N. Yamasaki, Portioned EDF-based scheduling on multiprocessors, in *EMSOFT*, 2008

70. A. Abel, F. Benz, J. Doerfert, B. Dörr, S. Hahn, F. Haupenthal, M. Jacobs, A.H. Moin, J. Reineke, B. Schommer, R. Wilhelm, Impact of resource sharing on performance and performance prediction: a survey, in *CONCUR*, 2013, pp. 25–43

71. C. Rochange, An overview of approaches towards the timing analysability of parallel architecture, in *PPES*, 2011, pp. 32–41

72. J.C. Palencia Gutiérrez, J.J. Gutiérrez García, M. González Harbour, On the schedulability analysis for distributed hard real-time systems, in *RTS*, 1997, pp. 136–143

73. K. Tindell, Holistic schedulability analysis for distributed hard real-time system. Micropro-cess. Microprogram. **40**(2–3), 117–134 (1994). doi:10.1016/0165-6074(94)90080-9. http://dx.doi.org/10.1016/0165-6074(94)90080-9

74. S. Chakraborty, S. Künzli, L. Thiele, A general framework for analysing system properties in platform-based embedded system designs, in *DATE*, 2003, pp. 10190–10195

75. J. Rox, R. Ernst, Compositional performance analysis with improved analysis techniques for obtaining viable end-to-end latencies in distributed embedded systems, in *STTT*, 2013

76. L. Thiele, S. Chakraborty, M. Naedele, Real-time calculus for scheduling hard real-time systems, in *ISCAS 2000*, vol. 4, 2000, pp. 101–104

77. J.-Y. Le Boudec, P. Thiran, *Network Calculus: A Theory of Deterministic Queuing Systems for the Internet* (Springer, Berlin/Heidelberg, 2001)

78. E. Wandeler, Modular performance analysis and interface-based design for embedded real-time systems. Ph.D. thesis, ETHZ, 2006

79. B. Jonsson, S. Perathoner, L. Thiele, W. Yi, Cyclic dependencies in modular performance analysis, in *EMSOFT*, 2008, pp. 179–188

80. E. Wandeler, L. Thiele, Real-Time Calculus (RTC) Toolbox (2006), http://www.mpa.ethz.ch/Rtctoolbox

81. J. Hennessy, D. Patterson, *Computer Architecture – A Quantitative Approach*, 4th edn. (Morgan Kaufmann, San Francisco, 2007)

82. H. Al-Zoubi, A. Milenkovic, M. Milenkovic, Performance evaluation of cache replacement policies for the spec cpu2000 benchmark suite, in *ACM-SE 42* (ACM, New York, NY, 2004), pp. 267–272

83. D. Eklov, N. Nikoleris, D. Black-Schaffer, E. Hagersten, Cache pirating: measuring the curse of the shared cache, in *ICPP* (IEEE Computer Society, Washington, DC, 2011), pp. 165–175

84. G. Grohoski, Niagara-2: a highly threaded server-on-a-chip, in *HotChips*, 2006

85. T.P. Baker, M. Cirinei, Brute-force determination of multiprocessor schedulability for sets of sporadic hard-deadline tasks, in *OPODIS*, 2007, pp. 62–75

86. Q. Deng, S. Gao, N. Guan, Z. Gu, G. Yu, Exact schedulability analysis for static-priority global multiprocessor scheduling using model-checking, in *Proceedings of the 5th IFIP Workshop on Software Technologies for Future Embedded and Ubiquitous Systems (SEUS)*, 2007

87. N. Guan, Z. Gu, M. Lv, Q. Deng, G. Yu, Schedulability analysis of global fixed-priority or edf multiprocessor scheduling with symbolic model-checking, in *ISORC*, 2008, pp. 556–560

88. Y. Li, V. Suhendra, Y. Liang, T. Mitra, A. Roychoudhury, Timing analysis of concurrent programs running on shared cache multi-cores, in *RTSS*, 2009, pp. 57–67

89. J. Yan, W. Zhang, Wcet analysis for multi-core processors with shared l2 instruction caches, in *RTAS*, 2008
90. W. Zhang, J. Yan, Accurately estimating worst-case execution time for multi-core processors with shared direct-mapped instruction caches, in *RTCSA*, 2009, pp. 455–463
91. S. Perathoner, EDF scheduling with real time calculus, in *Presentation*, TEC Group, Computer Engineering and Networks Laboratory, ETH Zurich, 2007
92. J.P. Erickson, J.H. Anderson, Fair lateness scheduling: reducing maximum lateness in g-edf-like scheduling, in *ECRTS* (2012)
93. H. Leontyev, J.H. Anderson, Generalized tardiness bounds for global multiprocessor scheduling. Real-Time Syst. **44**(1–3), 26–71 (2010)
94. M. Spuri, Analysis of deadline scheduled real-time systems, in *RR-2772, INRIA*, France, 1996
95. J. Reineke, Caches in wcet analysis – predictability, competitiveness, sensitivity. Ph.D. thesis, Saarland University, 2008
96. R. Heckmann, M. Langenbach, S. Thesing, R. Wilhelm, The influence of processor architecture on the design and the results of wcet tools, in *Proceedings of the IEEE* (2003)
97. A. Malamy, R.N. Patel, N.M. Hayes, Methods and apparatus for implementing a pseudo-lru cache memory replacement scheme with a locking feature. US Patent 5029072, 1994
98. A.S. Tanenbaum, *Modern Operating Systems*, 3rd edn. (Prentice Hall Press, Upper Saddle River, NJ, 2007)
99. D. Eklov, N. Nikoleris, D. Black-Schaffer, E. Hagersten, Cache pirating: measuring the curse of the shared cache, in *ICPP* (IEEE Computer Society, Washington, DC, 2011), pp. 165–175. doi:10.1109/ICPP.2011.15. http://dx.doi.org/10.1109/ICPP.2011.15
100. P. Kongetira, K. Aingaran, K. Olukotun, Niagara: a 32-way multithreaded sparc processor. IEEE Micro **25**(2), 21–29 (2005)
101. J. Reineke, D. Grund, Relative competitive analysis of cache replacement policies, in *Proceedings of the 2008 ACM SIGPLAN-SIGBED Conference on Languages, Compilers, and Tools for Embedded Systems, LCTES '08* (ACM, New York, NY, 2008), pp. 51–60
102. Y.-T.S. Li, S. Malik, A. Wolfe, Cache modeling for real-time software: beyond direct mapped instruction caches, in *RTSS* (IEEE Computer Society, Washington, DC, 1996), pp. 254–261
103. R.D. Arnold, F. Mueller, D.B. Whalley, M.G. Harmon, Bounding worst-case instruction cache performance, in *RTSS* (IEEE Computer Society, 1994), pp. 172–181
104. F. Mueller, Static cache simulation and its applications. Ph.D. thesis, Florida State University, 1994
105. F. Mueller, Timing analysis for instruction caches. Real-Time Syst. **18**(2/3), 217–247 (2000). doi:10.1023/A:1008145215849. http://dx.doi.org/10.1023/A:1008145215849
106. C. Ballabriga, H. Casse, Improving the first-miss computation in set-associative instruction caches, in *ECRTS* (IEEE Computer Society, Washington, DC, 2008), pp. 341–350
107. C. Cullmann, Cache persistence analysis: a novel approach theory and practice, in *LCTES* (ACM, New York, NY, 2011), pp. 121–130
108. C. Ferdinand, R. Wilhelm, On predicting data cache behavior for real-time systems, in *Proceedings of the ACM SIGPLAN Workshop on Languages, Compilers, and Tools for Embedded Systems, LCTES '98* (Springer, London, 1998), pp. 16–30
109. R. Sen, Y.N. Srikant, Wcet estimation for executables in the presence of data caches, in *Proceedings of the 7th ACM & IEEE International Conference on Embedded Software, EMSOFT '07* (ACM, New York, NY, 2007), pp. 203–212
110. B.K. Huynh, L. Ju, A. Roychoudhury, Scope-aware data cache analysis for wcet estimation, in *RTAS* (IEEE Computer Society, Washington, DC, 2011), pp. 203–212
111. D. Hardy, I. Puaut, Wcet analysis of multi-level non-inclusive set-associative instruction caches, in *RTSS* (IEEE Computer Society, Washington, DC, 2008), pp. 456–466
112. T. Sondag, H. Rajan, A more precise abstract domain for multi-level caches for tighter wcet analysis, in *RTSS* (IEEE Computer Society, Washington, DC, 2010), pp. 395–404
113. Y. Li, V. Suhendra, Y. Liang, T. Mitra, A. Roychoudhury, Timing analysis of concurrent programs running on shared cache multi-cores, in *RTSS* (IEEE Computer Society, Washington, DC, 2009), pp. 57–67

114. S. Chattopadhyay, A. Roychoudhury, T. Mitra, Modeling shared cache and bus in multi-cores for timing analysis, in *Proceedings of the 13th International Workshop on Software & Compilers for Embedded Systems, SCOPES '10* (ACM, New York, NY, 2010), pp. 6:1–6:10

115. J. Staschulat, R. Ernst, Scalable precision cache analysis for real-time software, in *ACM Transactions on Embedded Computing Systems*, vol. 6, no. 4 (ACM, New York, 2007). doi:10.1145/1274858.1274863. http://doi.acm.org/10.1145/1274858.1274863

116. S. Altmeyer, C. Maiza, J. Reineke, Resilience analysis: tightening the crpd bound for set-associative caches, in *LCTES* (ACM, New York, NY, 2010), pp. 153–162

117. J. Reineke, D. Grund, Sensitivity of cache replacement policies, in *ACM Transactions on Embedded Computing Systems*, vol. 12 (ACM, New York, 2013), pp. 42:1–42:18. doi:10.1145/2435227.2435238. http://doi.acm.org/10.1145/2435227.2435238

118. P.P. Puschner, B. Alan, Guest editorial: a review of worst-case execution-time analysis. Real-Time Syst. **18**(2/3), 115–128 (2000). http://dblp.uni-trier.de/db/journals/rts/rts18.html#PuschnerB00

119. F.E. Allen, Control flow analysis, in *Proceedings of Symposium on Compiler Optimization* (ACM, New York, NY, 1970), pp. 1–19

120. A.V. Aho, R. Sethi, J.D. Ullman, *Compilers: Principles, Techniques, and Tools* (Addison-Wesley Longman Publishing Co. Inc., Boston, MA, 1986)

121. J. Gustafsson, A. Betts, A. Ermedahl, B. Lisper, The mälardalen wcet benchmarks: past, present and future, in *10th International Workshop on Worst-Case Execution-Time Analysis, WCET'10*, 2010

122. X. Li, Y. Liang, T. Mitra, A. Roychoudhury, Chronos: a timing analyzer for embedded software. Sci. Comput. Program. **69**(1–3), 56–67 (2007). doi:10.1016/j.scico.2007.01.014. http://dx.doi.org/10.1016/j.scico.2007.01.014

123. T. Austin, E. Larson, D. Ernst, Simplescalar: an infrastructure for computer system modeling. Computer **35**(2), 59–67 (2002)

124. M. Lv, CATE: a simulator for cache analysis technique evaluation in WCET estimation (2012). Available from http://faculty.neu.edu.cn/ise/lvmingsong/cate/

125. M. Berkelaar, lp_solve: (mixed integer) linear programming problem solver (2003). Available from ftp://ftp.es.ele.tue.nl/pub/lp_solve

126. R. Wilhelm, Why AI + ILP is good for WCET, but MC is not, nor ILP alone, in *VMCAI*, 2004

127. E.W. Dijkstra, Chapter I: notes on structured programming, in *Structured Programming* (Academic, London, 1972), pp. 1–82

128. M. Berkelaar, lp_solve: a mixed integer linear program solver. Relatorio Tecnico, Eindhoven University of Technology, 1999

129. M. Bertogna, M. Cirinei, Response-time analysis for globally scheduled symmetric multiprocessor platforms, in *Proceedings of the 28th IEEE Real-Time Systems Symposium (RTSS)*, 2007

130. E. Bini, T.H.C. Nguyen, P. Richard, S.K. Baruah, A response-time bound in fixed-priority scheduling with arbitrary deadlines. IEEE Trans. Comput. **58**(2), 279–286 (2009)

131. K. Tindell, H. Hansson, A. Wellings, Analysing realtime communications: controller area network (can), in *RTSS*, 1994

132. A. Burns, A. Wellings, *Real-Time Systems and Programming Languages*, 3rd edn. (Addison-Wesley, Boston, 2001)

133. L. Lundberg, Multiprocessor scheduling of age constraint processes, in *RTCSA*, 1998

134. N. Guan, W. Yi, Z. Gu, Q. Deng, G. Yu, New schedulability test conditions for non-preemptive scheduling on multiprocessor platforms, in *RTSS*, 2008

135. T.P. Baker, A comparison of global and partitioned edf schedulability tests for multiprocessors. Technical Report, Department of Computer Science, Florida State University, FL, 2005

136. M. Bertogna, M. Cirinei, G. Lipari, Schedulability analysis of global scheduling algorithms on multiprocessor platforms. IEEE Trans. Parallel Distrib. Syst. **20**(4), 553–566 (2008). doi:10.1109/TPDS.2008.129. http://dx.doi.org/10.1109/TPDS.2008.129

137. K. Jeffay, D.F. Stanat, C.U. Martel, On non-preemptive scheduling of periodic and sporadic tasks, in *Proceedings of the 12th IEEE Real-Time Systems Symposium (RTSS)*, 1991

138. H. Leontyev, J.H. Anderson, A unified hard/soft real-time schedulability test for global edf multiprocessor scheduling, in *Proceedings of the 29th IEEE Real-Time Systems Symposium (RTSS)*, 2008

139. J. Goossens, S. Funk, S. Baruah, Priority-driven scheduling of periodic task systems on multiprocessors. Real-Time Syst. **25**(2–3), 187–205 (2003). doi:10.1023/A:1025120124771. http://dx.doi.org/10.1023/A:1025120124771

140. C.A. Phillips, C. Stein, E. Torng, J. Wein, Optimal time-critical scheduling via resource augmentation, in *Proceedings of the 29th Annual ACM Symposium on the Theory of Computing (STOC)*, 1997

141. S.K. Baruah, S. Chakraborty, Schedulability analysis of non-preemptive recurring real-time tasks, in *The 14th International Workshop on Parallel and Distributed Real-Time Systems (WPDRTS)*, 2006

142. S.K. Baruah, The non-preemptive scheduling of periodic tasks upon multiprocessors. Real-Time Syst. **32**(1–2), 9–20 (2006). doi:10.1007/s11241-006-4961-9. http://dx.doi.org/10.1007/s11241-006-4961-9

143. M. Blum, R.W. Floyd, V. Pratt, R.L. Rivest, R.E. Tarjan, Time bounds for selection. J. Comput. Syst. Sci. **7**(4), 448–461 (1973). doi:10.1016/S0022-0000(73)80033-9. http://dx.doi.org/10.1016/S0022-0000(73)80033-9

144. S.K. Baruah, A. Burns, Sustainable scheduling analysis, in *Proceedings of the 27th IEEE Real-Time Systems Symposium (RTSS)*, 2006

145. A.K. Mok, W.-C. Poon, Non-preemptive robustness under reduced system load, in *Proceedings of the 26th IEEE Real-Time Systems Symposium (RTSS)*, 2005

146. R. Ha, J.W.S. Liu, Validating timing constraints in multiprocessor and distributed real-time systems, in *Proceedings of the 14th International Conference on Distributed Computing Systems (ICDCS)*, 1994

147. T.P. Baker, S. Baruah, Sustainable multiprocessor scheduling of sporadic task systems, in *ECRTS*, 2009

148. J. Anderson, A. Srinivasan, Mixed pfair/erfair scheduling of asynchronous periodic tasks J. Comput. Syst. Sci. **68**(1), 157–204 (2004). doi:10.1016/j.jcss.2003.08.002. http://dx.doi.org/10.1016/j.jcss.2003.08.002

149. S.K. Baruah, N.K. Cohen, C.G. Plaxton, D.A. Varvel, Proportionate progress: A notion of fairness in resource allocation. Algorithmica **15**(6), 600–625 (1996). doi:10.1007/BF01940883. http://dx.doi.org/10.1007/BF01940883

150. U. Devi, J. Anderson, Tardiness bounds for global edf scheduling on a multiprocessor, in *IEEE Real-Time Systems Symposium (RTSS)*, 2005

151. K. Lakshmanan, R. Rajkumar, J. Lehoczky, Partitioned fixed-priority preemptive scheduling for multi-core processors, in *ECRTS*, 2009

152. B. Andersson, E. Tovar, Multiprocessor scheduling with few preemptions, in *RTCSA*, 2006

153. B. Andersson, K. Bletsas, Sporadic multiprocessor scheduling with few preemptions, in *Euromicro Conference on Real-Time Systems (ECRTS)*, 2008

154. S. Kato, N. Yamasaki, Real-time scheduling with task splitting on multiprocessors, in *IEEE Conference on Embedded and Real-Time Computing Systems and Applications (RTCSA)*, 2007

155. S. Kato, N. Yamasaki, Y. Ishikawa, Semi-partitioned scheduling of sporadic task systems on multiprocessors, in *ECRTS*, 2009

156. S. Kato,N. Yamasaki, Portioned static-priority scheduling on multiprocessors, in *IPDPS*, 2008

157. S. Kato, N. Yamasaki, Semi-partitioned fixed-priority scheduling on multiprocessors, in *RTAS*, 2009

158. J.W.S. Liu, *Real-Time Systems* (Prentice Hall, Upper Saddle River, 2000)

159. J.P. Lehoczky, L. Sha, Y. Ding, The rate monotonic scheduling algorithm: exact characterization and average case behavior, in *RTSS*, 1989

160. T. Baker, An analysis of EDF schedulability on a multiprocessor. IEEE Trans. Parallel Distrib. Syst. **16**(8), 760–768 (2005). doi:10.1109/TPDS.2005.88. http://doi.ieeecomputersociety.org/10.1109/TPDS.2005.88

161. H. Cho, B. Ravindran, E. Jensen, An optimal realtime scheduling algorithm for multiprocessors, in *RTSS*, 2006
162. K. Funaoka et al., Work-conserving optimal real-time scheduling on multiprocessors, in *ECRTS*, 2008
163. S. Lauzac, R. Melhem, D. Mosse, An efficient rms admission control and its application to multiprocessor scheduling, in *IPPS*, 1998
164. D. Chen, A.K. Mok, T.W. Kuo, Utilization bound revisited. IEEE Trans. Comput. **52**(3), 351–361 (2003). doi:http://doi.ieeecomputersociety.org/10.1109/TC.2003.1183949
165. B.K. Bershad, B.J. Chen, D. Lee, T.H. Romer, Avoiding conflict misses dynamically in large direct mapped caches, in *ASPLOS*, 1994
166. J. Herter, J. Reineke, R. Wilhelm, Cama: cache-aware memory allocation for wcet analysis, in *ECRTS*, 2008
167. J. Rosen, A. Andrei, P. Eles, Z. Peng, Bus access optimization for predictable implementation of real-time applications on multiprocessor systems-on-chip, in *RTSS*, 2007
168. A. Fedorova, M. Seltzer, C. Small, D. Nussbaum, Throughput-oriented scheduling on chip multithreading systems. Technical Report, Harvard University, 2005
169. D. Chandra, F. Guo, S. Kim, Y. Solihin, Predicting inter-thread cache contention on a multiprocessor architecture, in *HPCA*, 2005
170. J.H. Anderson, J.M. Calandrino, U.C. Devi, Real-time scheduling on multicore platforms, in *RTAS*, 2006
171. J.H. Anderson, J.M. Calandrino, Parallel real-time task scheduling on multicore platforms, in *RTSS*, 2006
172. J.M. Calandrino, J.H. Anderson, Cache-aware real-time scheduling on multicore platforms: heuristics and a case study, in *ECRTS*, 2008
173. K. Danne, M. Platzner, An edf schedulability test for periodic tasks on reconfigurable hardware devices, in *LCTES*, 2006
174. N. Guan, Q. Deng, Z. Gu, W. Xu, G. Yu, Schedulability analysis of preemptive and non-preemptive edf on partial runtime-reconfigurable fpgas, in *ACM Transaction on Design Automation of Electronic Systems*, vol. 13, no. 4 (2008)
175. N. Fisher, J. Anderson, S. Baruah, Task partitioning upon memory-constrained multiprocessors, in *RTCSA*, 2005, p. 1
176. V. Suhendra, C. Raghavan, T. Mitra, Integrated scratchpad memory optimization and task scheduling for mpsoc architectures, in *CASES*, 2006
177. H. Salamy, J. Ramanujam, A framework for task scheduling and memory partitioning for multi-processor system-on-chip, in *HiPEAC*, 2009
178. A. Wolfe, Software-based cache partitioning for real-time applications. J. Comput. Softw. Eng. **2**(3), 315–327 (1994). http://dl.acm.org/citation.cfm?id=200781.200792
179. S. Dropsho, C. Weems, Comparing caching techniques for multitasking real-time systems. Technical Report, University of Massachusetts-Amherst, 1997
180. B.D. Bui, M. Caccamo, L. Sha, J. Martinez, Impact of cache partitioning on multi-tasking real time embedded systems, in *RTCSA*, 2008
181. D. Chiou, S. Devadas, L. Rudolph, B.S. Ang, Dynamic cache partitioning via columnization. Technical Report, MIT, 1999
182. D. Tam, R. Azimi, M. Stumm, L. Soares, Managing shared l2 caches on multicore systems in software, *WIOSCA*, 2007
183. J. Liedtke, H. Hartig, M. Hohmuth, Os-controlled cache predictability for real-time systems, in *RTAS*, 1997
184. N. Guan, M. Stigge, W. Yi, G. Yu, Cache-aware scheduling and analysis for multicores. Technical Report, Uppsala University, (http://user.it.uu.se/yi), 2009
185. C. Kim, D. Burger, S.W. Keckler, An adaptive, nonuniform cache structure for wiredelay dominated on-chip caches, in *ASPLOS*, 2002
186. L. Thiele, S. Chakraborty, M. Naedele, Real-time calculus for scheduling hard real-time systems, in *Proceedings of the International Symposium on Circuits and Systems*, 2000

187. J.L. Boudec, P. Thiran, Network calculus – a theory of deterministic queuing systems for the internet, in *LNCS 2050* (Springer, Berlin, 2001)
188. A. Maxiaguine, Y. Zhu, S. Chakraborty, W.-F. Wong, Tuning soc platforms for multimedia processing: identifying limits and tradeoffs, in *CODES+ISSS*, 2004
189. J. Bengtsson, W. Yi, Timed automata: semantics, algorithms and tools, in *Lectures on Concurrency and Petri Nets*, 2003
190. S. Perathoner, T. Rein, L. Thiele, K. Lampka, J. Rox, Modeling structured event streams in system level performance analysis, in *LCTES*, 2010
191. L. Thiele E. Wandeler, Characterizing workload correlations in multi processor hard real-time systems, in *RTAS*, 2005
192. L. Thiele, E. Wandeler, N. Stoimenov, Real-time interfaces for composing real-time systems, in *EMSOFT*, 2006
193. L.T.X. Phan, S. Chakraborty, P.S. Thiagarajan, A multi-mode real-time calculus, in *RTSS*, 2008
194. K. Tindell, J. Clark, Holistic schedulability analysis for distributed hard real-time systems. Microprocess. Microprogram. **40**(2–3), 117–134 (1994). doi:10.1016/0165-6074(94)90080-9. http://dx.doi.org/10.1016/0165-6074(94)90080-9
195. M. Gonzalez Harbour, J.J. Gutierrez Garcia, J.C. Palencia Gutierrez, J.M. Drake Moyano, Mast: modeling and analysis suite for real time applications, in *ECRTS* (2001)
196. S. Perathoner, E. Wandeler, L. Thiele, A. Hamann, S. Schliecker, R. Henia, R. Racu, R. Ernst, M. Gonzalez Harbour, Influence of different abstractions on the performance analysis of distributed hard real-time systems. Des. Autom. Embed. Syst. **13**(1–2), 27–49 (2009). doi:10.1007/s10617-008-9015-1. http://dx.doi.org/10.1007/s10617-008-9015-1
197. K. Lampka, S. Perathoner, L. Thiele, Analytic real-time analysis and timed automata: a hybrid method for analyzing embedded real-time systems, in *EMSOFT*, 2009
198. M. Moy, K. Altisen, Arrival curves for real-time calculus: the causality problem and its solutions, in *TACAS*, 2010
199. S.K. Baruah, D. Chen, S. Gorinsky, A.K. Mok, Generalized multiframe tasks. Real-Time Syst. **17**(1), 5–22 (1999). doi:10.1023/A:1008030427220. http://dx.doi.org/10.1023/A:1008030427220
200. K. Richter, Compositional scheduling analysis using standard event models. Ph.D. thesis, Technical University Carolo-Wilhelmina of Braunschweig, 2005
201. K. Richter, Compositional scheduling analysis using standard event models. Ph.D. thesis, Technical University of Braunschweig, 2004
202. V. Pollex, S. Kollmann, F. Slomka, Generalizing response-time analysis, in *RTCSA*, 2010
203. L. Thiele et al., A framework for evaluating design tradeoffs in packet processing architectures, in *DAC*, 2002
204. A. Easwaran, M. Anand, I. Lee, Compositional analysis framework using edp resource models, in *RTSS*, 2007
205. I. Shin, I. Lee, Compositional real-time scheduling framework, in *RTSS*, 2004
206. C. Kenna, J. Herman, B. Brandenburg, A. Mills, J. Anderson, Soft real-time on multiprocessors: are analysis-based schedulers really worth it?, in *RTSS*, 2011
207. S. Perathoner, EDF scheduling with real time calculus, in *Presentation Slides*, TEC Group, Computer Engineering and Networks Laboratory, ETH Zurich, 2007
208. E. Bini, G. Buttazzo, The space of EDF deadlines: the exact region and a convex approximation. Real-Time Syst. **41**(1), 27–51 (2009). doi:10.1007/s11241-008-9060-7. http://dx.doi.org/10.1007/s11241-008-9060-7